The First 100:

Ideas & Interpretations

By
Charles K. Summers

https://technoglot.blogspot.com

The First 100: Ideas & Interpretations by Charles K. Summers

Library of Congress Control Number: 2020908234

ISBN-13: 978-1-7349129-0-6

1) Technology 2) Reference 3) Body & Mind

First Edition

Table of Contents

Introduction

I am not an expert in all topics – I don't even consider myself an expert in ANY topic. For any expert in a topic, there is usually an even greater expert in that topic. Thus, although I have researched to the best of my time and abilities – I do not claim to be "right" in my presentations. These are my thoughts, as expressed in my blogs over an eleven-year period – from 2007 through 2017.

I was born as the son of a waitress and a roughneck (a roughneck is actually someone who does the dirty work on oil rigs) in a small rural town in Kansas. I moved between different towns in Kansas and ended up graduating from high school in Lewiston, Idaho.

Always curious about just about everything, my initial majors in college were in electrical engineering, math, and French. My last degree was a Master's in Computer Science. I started work as a programmer with the Agricultural Department at Kansas State University but my first full-time job was with Bell Laboratories in Westminster, Colorado. Later, a friend and I started a company as a "virtual corporation" -- everyone working SOHO -- where I was the VP of Engineering (of a small company). Presently, I work as a Programming Manager for a very large networking/security company.

These blogs started off (as the first journal entry details) as a technology speaking journal (technoglot) – to explain daily technology to those who haven't been trained in such areas. As time passed, the topics have tended to vary more – with a definite leaning towards economic aspects since much of our worldview tends to be molded by economic

structure. If you prefer to not buy this, you can read the original blog posts directly at
https://technoglot.blogspot.com

I hope that these are of interest to you and arouse a desire to talk among your friends and continue to do research on the areas I have mentioned.

Charles K. Summers
May 2020

1. What is a Technoglot - 9/26/2007

I am a geek, a dweeb, and a nerd. On the other hand, I have lived almost 50 years in the land of the "normal" and I have survived and have even thrived at times. I also started off with a love of languages -- studying French in high school, continuing with German and Russian in college or after, and working on Farsi and Spanish nowadays. In high school and college, I learned the basics (or more) of about 14 computer languages from assembler to JCL to COBOL to FORTRAN to LISP to C. I have also been involved with music (violin and voice), art (lithography and drawing), and "crafts" (which can also attain the level of art, in my opinion) such as weaving and woodworking.

Language shapes thought and thought shapes language. This is true of all languages, be they "natural" languages, computer languages, or the languages of music, art, and craft. The inclination of Russian speakers to enter into the passive form or the propensity of the French to fall into allusion and metaphor reflect (or mold?) the way they are perceived and interact with the world. A programmer starting from the object-oriented world will look at problems from an object-oriented viewpoint as someone starting from a "structured" language environment will have their own orientation.

And here we start to show the differences between languages. What does "object-oriented" mean? How does it differ from "structured"? What is allusion? What is passive? Even when we believe that we are speaking in the same language, our history, education, experience, and environment changes what we actually hear and understand. An Inuit may have more than a dozen words for "snow" and

1

Greek may have five (or more) words for "love" but in order to translate that to a language which does **not** have those words you end up with a definition that is not always able to be used directly within the context of a sentence or thought. English has grown to be the most widely spoken (not necessarily first language) language in the world by its extreme inclusionistic attitude; if you don't have the word in English then adopt it.

Communication is the process of producing information, transporting it, and understanding it. This happens with speech which is created, spoken, and heard. Or with sign language which is created, formed, and seen. It also happens between a user and a computer which is essentially a task of translation from one medium (screen display, keyboard, mouse) and language (symbol, letter, position) to/from storage and analysis. It also happens directly between computers in a method of communication protocols.

Protocols consist of the syntax and semantics of a language. I have worked with protocols most of my life -- from learning English as a baby, to learning to program, to more than 20 years of design and implementation of protocols allowing both data and signaling communication between microprocessors (or computers, if you prefer). In this area, I have published books on Integrated Services Digital Network (ISDN) (see "ISDN Implementor's Guide" at amazon.com) and Asymmetric Digital Subscriber Line (ADSL) architectures ("ADSL: Standards, Implementation, and Architecture") which help to explain how these protocols work and can be used. From my work on these, and dozens of other data, transport, and signaling protocols, I have a sound foundation for understanding and defining other protocols that may not even yet be created.

Yet computer-to-computer protocols are still "easy". Why? Because the same definitions are used for the sender (transmitter) and receiver. If communication fails, it is because something was not implemented correctly or the definition of the protocol was faulty. So, it is altered until it does work.

But the same is not true for user-to-computer and people-to-people communication. Why? Because people do not use the same language as other people or computers. Even people who are both, in theory, speaking the same language really are not. This is because their understanding (as opposed to hearing, seeing, or feeling) of the language is based on their own viewpoint which, in turn, is created by their history, education, and environment (physical, societal, and cultural).

So, back to the beginning and the title. What is a technoglot? It is someone who "speaks tech". In these blogs, I will strive to help understanding of technology for the non-technologist and to communicate ideas from one language to another. I hope that you have fun and that my blogs are useful.

2. What Makes a Phone Mobile?
10/5/2007

I have a mobile phone. No, it isn't a cellular phone. It's a cordless phone and I love it. I can work on the laundry or putter about in the kitchen or even go out and get the mail and still be able to converse on the phone.

But, I admit, that's not what most people think of when you say a mobile phone. So, what is?

A phone has basically two parts. One part connects to the network, or phone company, or central office, or carrier, or whatever term you want to use. Over that part, the information related to how, and to whom, you want to connect to someone else is conveyed. There are a lot of different mechanisms from POTS (Plain Old Telephone System), to ADSL (sometimes the "A" is dropped nowadays), to ISDN, to optical fiber using a version of ATM over SONET. What do all those mean? Not much to most people although it is important to the service provider and makes a difference as to what things you can do over your phone. The main point is that it lets you be called and make calls.

The other part is the control part. Usually that means a keypad but it may be voice activated or have a touch pad. This lets you enter in the information needed to connect to whom you want to connect -- or make use of the services provided (call waiting, Instant Messaging, etc.).

The control part can be all in one unit -- traditional phone or "cell phone". Or, as in the case of my cordless phone, it can be separated from the control base and the information exchanged very locally (usually less than 500 feet).

The point of a mobile phone is mobility. Presently, that means more than 500 feet to most people. What do you need for that?

Basically, you have to extend your "access point". The access point is where you connect to the network. A "landbased" line has a physical line that connects to a jack in the wall (well, there are some other ways -- but that's the way almost all are connected). A "cellular" phone connects via "airwaves" (let's leave out the various frequency bands for the moment) to a transmission tower which is the access point.

And that is what is meant by a mobile phone -- extending the access point. How is that doe? Stay tuned for that in the next post.

3. From Cordless to Mobile – 8/10/2009

Apologies for taking so long to resume my posts. Frankly, it was a bit difficult to feel like posting when it seemed like no one was paying any attention. But, I've decided that that is no reason not to post, and I really appreciate the feedback from my one (perhaps only) reader -- thank you Leda. Please, everyone, do tell me about what you are interested.

Last post, I talked about going into the details of what makes a mobile/cellular phone different from a cordless phone. There really isn't that much real difference. The primary differences are distance between the handset and the base station and the fact that each base (cell) station will handle more than one phone. In addition, there is a need for handover (handoff) of the phone from one base (cell) station to another as one moves around.

When one uses a cordless phone, it must synchronize with the base station. To do this, it may require cycling between "channels" to find a clear connection. However, it is still a relatively simple connection -- usually no need for identification or security.

A cell phone goes through a number of stages. When it is powered up, it basically "shouts" into the air and these transmissions are omnidirectional (unlike a satellite link, the handset may be in any position relative to the base station) so the signal must go all directions. Although it is possible for carriers to share a pool of transmission frequencies, it is more common for a single carrier to use a set of frequencies. So, the phone shouts out within this bandwidth and waits for a response.

The cell station listens and then, using identification information found in the initial transmission from the mobile handset, sends back a response. It is possible that a further exchange of information (called a handshake) will take place to further identify the specific phone and the services available to that handset. This process is known as registering. Much of the registration information will reside in the memory (or specialized memory -- known as a SIM) of the handset. The first time you use your handset, much more information is exchanged to build up the store of data in the handset.

Once registered, people can call you and you can call others. Signal strength (or number of "bars") will make a difference about the clarity of the call. However, calls are really made between the base station and the other phone -- not the handset. The handset is the final destination/origination of the data (voice or email or whatever) but the call is really between the base station and the other end (peer).

The base station may make an additional jump to a more powerful station or it may directly connect into the wired "landline" phone networks. At some point, almost all mobile calls are actually routed via landlines. It is just more efficient to use the Radio Frequency (RF).

The ability to do the "handover" is what really distinguishes a mobile phone from a cordless phone. As you move, with your handset, the signal strength from the current base (cell) station will change. If it decreases, it may reach the point where the signal just isn't strong enough to continue the call. When the signal strength decreases, the handset will start shouting again. If it finds a new, stronger, base station, it will initiate the "handover", which is basically the same thing as a "transfer" from a non-mobile phone. The call is transferred from the old base station to the new base station and it will then act as the primary endpoint for the call (with data continuing to be exchanged with the handset).

This transition must be done very quickly to allow an interrupted use of the circuit. Because of this transition delay, the handset and base station may behave a bit differently if a data session is in use -- perhaps setting a lower signal strength trigger for handover and waiting for a pause in data to do the transfer.

Of course, all this doesn't actually work as simply as described. In particular, the RF transmission and reception are very difficult areas to engineer. Although I have taken enough physics and engineering courses to understand, in principle, how the RF works, my general impression is that RF engineers are magicians.

So, that is really the difference between a cordless phone and a mobile phone -- the ability to do that transfer (handover) as smoothly as possible.

I leave you at that point but will try to post on a much more regular basis.

4. It's Not Easy to Drive Green – 8/11/2009

As the joke goes, the simplest way to drive a green car is to take it into the body shop for a new paint job. Personal automotive transport just isn't the overall best way to reduce carbon output. Walk, bicycle, ride public transportation, skateboards, pogo sticks -- fill it in.

But folks love having their own "wheels" -- the freedom to go when and where they want and not have to take a long time in the process. So, recognizing that it'll never be the "best" solution, what are the ways to make it as green as possible?

First thing to do is to recognize that there are two different statistics involved with the eco-friendliness of transportation. These are the local carbon footprint and the overall effects on the environment.

On the local carbon footprint, matters are fairly clear. If you use a traditional gas/diesel engine, get one that has the best MPG. Next down comes hybrid cars -- which have an "extra" in that they pollute less when stopped at an intersection. Yet cleaner (locally), are natural gas, all electric, hydrogen, and solar (not quite feasible yet -- but perhaps soon).

So, if everyone drives an electric car, there will be no more pollution and the world will be saved. Right?

No, not that simple. Of course, there is the fact that private transportation is only part (albeit an important part) of the climate problem. But the big problem is that it doesn't cover the overall effects on the environment -- which are NOT easy to determine and which are not always immediately "obvious".

The June 2008 issue of Wired mentions that manufacture of batteries for hybrid cars has a BIG carbon footprint. I don't have the issue nearby but it offsets the carbon savings from using a hybrid by many years (5 to 10, as I recall). So, a Smart car is still a better bet -- in spite of the excitement of hybrid.

What about the zero-emission alternatives? Well, they have to be manufactured. They also need roads to drive on. They need parking spaces (take a peek from Google Earth and just see how much of the land is occupied by roads and parking spaces -- especially in cities). And, except for yet-to-be-viable solar cars, their fuel still needs to created and transported.

Let's say that you have an all-electric car. How is that electricity produced? Is it a solar farm or a geothermal plant? Well, probably not (though possible). The electricity is produced and the conversion of energy is not 100% efficient (it varies a lot depending on design and fuel). Wikipedia indicates that there is a little over 7% loss in transporting over the electrical grid.

Depending on many factors -- such as how eco-conscientious the power producer may be, that electric car may actually have a larger carbon footprint than a Smart car. (I keep using a Smart car as an example -- if you prefer, just exchange that for a "very small, efficient internal combustion car".)

I used electric cars as an example but the same analysis holds for hydrogen cars.

The bottom line is that it's not easy to drive green -- so check the first paragraph for truly better alternatives. Then, if that doesn't fit with your life, organize your trips to reduce miles. Get as small of a car as works for you -- rent something for those "once a year" situations. Live and work locally if at all possible. Reduce the miles driven and spend more time with your friends and family.

Most of all, keep the "big picture" in mind. Don't feel guilty -- act responsibly.

5. Smart Plugs, Dumb Plugs and Dumb Designs – 8/13/2009

I read a lot of emails, and articles in the magazines I read, about the next way to save electricity.

Some of these are related to the problem of passive power loss. This occurs when an electrical appliance is plugged in but is not actively in use. If designed well, then they will not use any power when they are turned off -- but what is off?

Any device that can be turned on by a remote is not completely without power. Any device that maintains an active clock (not powered by batteries) is not completely without power. And, in today's world, that makes up a large percentage of the electrical devices in our houses.

A "smart" electrical strip is designed to have one unit on a power monitoring circuit. The remaining outlets are triggered to provide power only if the monitored outlet starts consuming electricity beyond a certain level. For example, a television might be plugged into the monitored outlet. When it is "off", it only requires a certain level of power but, when on, it greatly increases the electricity needed. At this point, the other outlets become active. A DVD player, for example, might be plugged into a "client" outlet and the television into the monitored outlet. Until the television is turned on, no power is given to the DVD player (and the remote will not work with it).

Unfortunately, this is very limited in its uses.
A DVR cannot be plugged into a client outlet because it runs timed events (an active clock). Same is true with a VCR. It

MIGHT be useful with an amplifier or receiver but not with a satellite receiver which requires passive checks for updates of data over the phone lines. Every potential scenario must be carefully examined.

I purchased a "Kill-A-Watt" power monitor via a "Gold sale" on amazon.com. It seems to be designed well (with a few not-too-important caveats) but I have been unsuccessful in tracking down the largest electrical uses within the house. I thought that my study -- with four computers, a printer, LAN, and quite a few other individual devices (such as hard drives and powered disk readers) was surely the culprit. But it wasn't -- so the "Kill-A-Watt" was useful to disprove that. I now believe that the main drains are the pool pump and the house fans -- usually direct wired and often badly inefficient and expensive to replace.

So, how to reduce electricity use? Unplug those items not in use (some DO use power even when "off"). Run as many items though an Uninterruptible Power Supply (UPS) as possible in a study -- a UPS will act as a buffer for items on reduced power demand. Reduce the number of internal clock devices. Keep the lights off when not in use. Use fans that have energy star ratings if possible -- and keep energy use in mind when purchasing new items.

6. Computer Literacy? – 8/14/2009

My 10-year-old, fifth grade, son came to me the other day -- he had a micro SD disk that he said had cheat codes on it that he wanted to use for some Wii games. I connected up my card reader to the USB connection on the Windows machine and looked at the files. I looked -- and then I rewatched the Youtube tutorial on how the program was supposed to work -- then I said to him "OK, this is the data file -- where is the program?" He responded "what is a program?".

Admittedly, only people with a historical bent are truly interested in how I started programming with little toggle switches on the front of a DEC computer -- or the huge decks of punched cards that I used in college to meet the needs of my programming classes. But, I do think that children who are in a school with a computer lab and multiple computers in the classroom should not be asking "what is a program?" by the time they're in fifth grade. And it's a good school.

Computer literacy seems to be interpreted as the ability to use particular programs (even if they don't know what a program is). My son did a couple of PowerPoint presentations last year for class and he's used Microsoft Word for a few papers. He has no problem in using a browser (even if he doesn't know what a browser is) and searching for information with Google. BUT, I still maintain that it is hard to transfer skills from one program to another without knowing some computer basics.

Many of the readers of this blog will already know most of this -- but I'm going to use my next week of blogs to talk about computer basics that I, personally, think ought to be known to be computer literate.

Subjects will include:

What is a Computer?

What is memory, files, and storage?

What are data (besides being a character on Star Trek: The Next Generation)?

What is an operating system (all my children keep complaining when a Windows program won't work on the family Macintosh)?

What are programs and how do they work?

What are peripherals -- what is I/O?

And any other topic that comes up within this realm. I once wrote a book (never published) on computers and how they work -- let's see if I can summarize in a set of reasonably sized blogs.

7. Computer Literacy 101 – What is a Computer? – 8/19/2009

Back from jury duty (actually, jury selection -- I didn't have to serve). I am going to start my short series on areas that I feel are important to become computer literate. I will try to make it useful but not to go into greater detail than necessary.

A computer, of today, is based on a semiconductor chip called a microprocessor. This is also sometimes referred to as a CPU (Central Processing Unit). This microprocessor changes numbers. It does very simple things with the numbers -- moves them from one location to another, shifts them (which effectively multiplies them by 2 or divides them by 2), does "logic" operations on them (which, if you don't know what that is -- will be another topic in the future), and changes the order in which it follows directions depending on the values that it has been given or has created based on the instructions given. That's about it, although some microprocessors will have other instructions that are actually shortcuts for a sequence of operations that **could** be done with a long series of simpler instructions.

The microprocessor does simple things with the numbers -- but it does them **very**, **very** quickly. When you see an advertisement for a "4 GHz CPU", it is an indication of how fast the microprocessor is and an indication of how quickly instructions can be processed. Larger numbers indicate a greater potential for speed. I say potential because the final speed depends on many design factors but, in general, you can use these numbers as an indicator of speed.

Microprocessors are used in many devices in our modern world. When they are in something that allows its use for different, non predetermined, purposes, it is called a "general purpose computer" or just plain old **computer**. Most people are familiar with smaller computers that are referred to as Personal Computers (PCs) but computers come in many sizes from a "smart" cell phone to a large "mainframe" which might be as large as several filing cabinets (they keep getting smaller, though).

The other category of use of microprocessors is for specialized uses -- the microprocessor is used for only one thing, or for one category of uses. Televisions, amplifiers, DVD players, watches, digital clocks, refrigerators, coffee makers, fuel injection and engine monitoring systems, traffic lights, alarm systems and MP3 players are all examples of devices using specialized processors. Some devices start entering into a "grey" area where they **could** be used for various, non-predetermined, purposes but are actually primarily used for a single purpose.

The microprocessor does not exist just by itself. It must have access to ways of giving it information upon which to work and ways to give results. In order to do this, other electronic devices will be connected to it and they, in turn, will be connected to other devices. A digital watch is an example of a very "simple" system which has buttons for setting information and a display to give results.

One very important additional item to which the microprocessor must have access is *memory*, which will be the topic of the next blog.

8. Computer Literacy 101 – What is Memory? – 8/20/2009

We talked about the microprocessor, or CPU, of a computer last blog -- but what can it do without someplace to store, and retrieve, information (next entry will talk about how that information -- or **data** -- is used)?

The first thing to know about memory for computers is the idea of address space. Address space serves the same purpose for computers as postal addresses do for paper mail. Often that address space will start at zero (0) and continue to as large as the physical memory allows. As an example, you have a 4GB flash drive that you use for assignments -- carried between home and work, or classes. The addresses of the information on that flash drive will be from 0 up to 4 gigabyte - 1 (or actually 4,294,967,295 but you can think of it as 4,000,000,000 units of information).

You also may have many different locations for memory -- these are called **storage** units. So, you may have a hard drive, a DVD-ROM drive, a USB flash card, and a set of backup files on a tape unit. Using the comparison to the paper mail system, let's say that each of these are in a different city -- so addresses consist of the information location (0 to 4 gigabyte on the flash drive) and storage location "G:flash_drive".

Alright, this is pretty cumbersome and it can really strain a microprocessor to directly address 4 gigabyte of memory (although it is possible for computers that describe themselves as "64-bit processors"). So, what most computer systems do is to break this memory down into separate "blocks" (or streets in the mail system). Each block may consist of 512 bytes. Therefore, we can now address a block as "G:flash_drive", "block 0 to 8,388,607" and a specific location as "G:flash_drive", "block 1,567", "byte 93".

OK. We see that by numbering the information and then breaking it down into separate groups we can effectively find all the information. But it is still rather clumsy talking about "G:flash_drive", "block 1,567", "byte 93". That is where the **file system** comes into play. A file system starts putting all of those numbers into friendlier terms. "Charles resume" is a lot easier to remember when one is doing a word processing program than telling it -- give me "C:CKS_hard_drive", "block 7,432 through 7,987", "bytes 0 (of first block) through 345 (of last block)".

Not only that, but this specific labeling of the location makes it very hard to move it -- what if I add something to the information (or **file**)? It will get bigger and the location will also get bigger -- do I now have to remember a whole new set of numbers? No. The file system will keep track of just where the file is now located and how large it is. (It also means that the file does not have to be *contiguous* -- the bytes can be located in different blocks and not come one after another.)

There are two other regions of memory that should be understood -- these are RAM and registers. Random Access Memory (RAM) is like your own pile of papers on your desk. It can be gotten to very quickly and is "right there". Registers are like "post-its" and are used for temporary copies of numbers to be used. RAM will often be addressed just by the byte numbers and registers will be referred to by special names (R1 for register 1) -- but these details aren't really important to most people.

Now that we have information in memory, how do we use it? That is the concept of data.

9. Computer Literacy 101 – What are Data? – 8/27/2009

All computers work with **data**. But, what are data? I say "what *are* data" because the word data is a plural one -- it is the plural of **datum**. However, almost no one ever uses the word datum and just about everyone treats data, grammatically, as singular. A datum is a single piece of information -- yes or no, it is raining or it is not raining, you have eaten breakfast or you have not eaten breakfast. You will note that a datum only indicates a yes/no or off/on, **binary**, condition. Most of the time, when we need information, it is really a collection of datum -- or data.

The same is true with computers -- they work on each individual datum but they pull them out of a pool of data. This data (I will use the conventional singular grammar here) is kept in storage, as I pointed out in a previous blog. It is then transferred from one storage area to the local RAM where the microprocessor can directly work with it.

Data is used by the microprocessor at many stages. The first stage, or startup (or **bootup**), is when the microprocessor first receives electricity. The actual hardware (the collection of semiconductor chips, and other discrete electronic components) is designed to start transferring data from a specific memory storage area and address. Often, this is address zero (0). This means that the microprocessor will transfer data from address 0 (the actual physical location, once again, depends on design of the hardware) to its working memory. It then **executes** the data -- it starts to perform specific operations based on the contents of that data that was at that address and then increments the address for the next instruction (usually by one -- unless the first instruction says something else) and then executes the operations for that address and on and on.

The next stage occurs once the **registers** (we talked about them in the storage blog) of the microprocessor have been filled with working data. At this stage, it is prepared to continue to execute instructions as it transfers them from memory. You can look at it as having two stages -- the first where the microprocessor "wakes up" with no specific contents in its registers and the second when the microprocessor has been **initialized** and can now proceed to work as the data tells it. Or, you can look at it as having three stages -- the first one the boot stage, the second is the startup stage where it is still getting ALL of the hardware connected to the computer ready to be used, and the third stage where any type of instructions can be executed because all of the hardware has been set up to be ready for use.

All of this works with the data. The data, for a general purpose microprocessor, is what makes it able to work differently each time it is turned on. For a specialized microprocessor, the incoming data starts the activities for which the microprocessor is designed.

Somehow, I managed to avoid the word **program** in this description but data are often split into two categories. Instruction data, or programs, are executable -- they contain instructions for the microprocessor while program data is used by the programs to produce more data. The difference between these categories of data is that instruction data does not change (unless someone specifically writes another program to change that instruction data -- the topic of **viruses** and software **patches).**

I'm going to swap the next two items on my original "computer literacy" list and talk about programs more in the next blog.

10. Computer Literacy 101 – What are Programs? – 8/28/2009

Data falls into two categories, as we saw in the previous blog. These categories are instruction data and program data. Instruction data can also be called a program -- which makes use of the program data to fulfill its purpose. Many times, a program will be called an **application**. Most people refer to programs as applications if they are widely used by different people. A word processing program may be considered an application. A spreadsheet program may be considered an application. Programs that people do not directly make use of are usually not called applications.

Programs consist of a sequence of instructions that tell the computer what to do. In the first blog of this series, we mentioned the types of instructions that may be given a microprocessor. These instructions are called **machine language** because they are sets of datum that are interpreted directly by the microprocessor. For example, the **decimal** value of 026002 will tell an HP 2100 computer to "jump" (change the current address for the next address) to address 002. The 026 portion is a "change the current instruction address to" code for the microprocessor and the 002 portion tells it the new value for the instruction address. Frankly, you don't really want to know much more than that unless you are involved with microprocessor design. Most microprocessors have their instructions written in **binary** (0s and 1s) or **hexadecimal** (symbols 1 through F for each numerical location) but the computers of the "early days" were not always standardized.

There is a hierarchy of languages used to program computers. At the base is machine language -- normally represented by a series of 0s and 1s -- like 10011001100011100011111101011110 -- which would be considered 32-**bit**s. Current-day programmers almost never use machine language directly. The next level is called **assembly language** which uses a set of readable codes that can be directly translated into machine language. An example of assembly language might be "JMP START", where JMP is the operation to be performed and START is a symbol for an address that is the value to be used by the operation. The next level consists of many different languages in a family called **high-level** computer languages. A **compiler** changes the high-level language into assembly language (or, sometimes, directly into machine language. An **assembler** changes the assembly language into machine language. Finally (at least, at this current point of time) there are **machine-independent** languages which are used to create programs that may be run on many different computers without being changed.

Programmers write programs. Very few write machine language programs. More, but not many, write assembly language programs. Most write programs in high-level languages. An increasing number write programs in machine-independent languages (such as **Java**). However, all of them end up actually creating machine language -- with special programs such as Java interpreters/compilers, compilers, and assemblers acting to make it into this special, final, form.

I said earlier that some programs are visibly used by people -- and these are called applications. The ones that are NOT visibly used by people are sometimes called **system programs**. These are programs that enable to computer to perform the acts that people want. A printer program will be used to allow people to print a document from their application. At the core of all of the system programs is a particular program called an **operating system** and this will be addressed in the next blog.

11. Computer Literacy 101 – What is an Operating System – 9/1/2009

An **operating system** (or OS) is a special program that allows other programs to run. This is the core of a general purpose computer -- the ability to run programs that are not pre-determined. A specialized computer (or microprocessor) may be able to run multiple programs but they are known in advance and, thus, the system can be designed to just run those programs (simpler, faster, and smaller).

As a more general definition, an operating system manages the resources of a computer. Sometimes, it is defined by a hierarchical model (sometimes known as the "onion" model) because there are services that are provided by the primary section of the operating system. Then there is another layer that provides new services plus makes use of the primary section. Then another layer that makes use of the other two layers, and so forth.

Some of the services provided by an operating system (in "layer order") are task (program) management, memory management, process (tasks communicating with one another) management, device management (such as hard disks, or CD-ROM drives, and so forth), and finally file management.

Because the applications make use of operating system services, which in turn rely on other "lower layer" services, most applications are not portable between different operating systems. This is why a program that works with Windows (a particular operating system) may not work on a Macintosh (running Mac OS X, a different specific operating system) or under a system running UNIX or Linux (two closely related operating systems).

There is an apparent exception to this -- but it really still follows the same rules. If one uses a browser that executes a program within the browser, that program may work with browsers running on computers using different OSs. It *appears* to be OS independent (able to work under any operating system) but, actually, it is making use of a set of services that have been defined to have the same use on multiple operating systems. Because this set of services, or **interface,** between the program and the next layer is the same, the program can run under different operating systems -- but the underlying program that provides that interface is still operating system dependent.

There is little agreement on what Operating System is "best". In general, the one that provides the services that you need, and the applications you need, for the lowest price and greatest speed is "best" for you. (It may not be the best for someone else with different needs.)

12. Computer Literacy 101 – What is a Peripheral? – 9/2/2009

A computer system cannot often stand alone -- it needs a way to input data, it needs a way to output data and it needs a way to expand its capabilities. These **devices** are called peripherals. A keyboard is a peripheral. A monitor is a peripheral. The time panel on a microwave oven can properly be called a peripheral, although it may be part of the overall design and not optional.

Peripherals can be grouped in classes. One class is Input/Output (or I/O). These peripherals allow you to put in (input) data or to access (output) data. Another class would include removable storage systems -- a flash card, a hard disk, a CD-ROM unit. Other classes exist.

Input devices have a particular requirement from the operating system. Since it is unknown just WHEN data will be input, there must be a way for the operating system to notice that data are available. The two main ways of doing this are via **polling** and via **interrupts**. A poll is a periodic check for data -- like a child in a car asking "Are we there yet?". An interrupt is like a tap on the shoulder. Different systems will use different methods. A poll does not take long but there will be many times that the answer will be "no" -- and, thus, the time taken is "wasted". An interrupt takes much more time because it is necessary to save the current situation (maybe you're in the middle of a program) before the interrupt can be handled. Let's say that handling an interrupt takes 50 milliseconds and a poll takes 1 millisecond. If an event occurs once a minute, then doing an interrupt will take less time than polling once a second (50 +

(60 x 1)). However, if the event occurs twice a minute, then polling is more efficient (50 + (30 x 1)).

A keyboard is an input peripheral. So is a mouse. In the case of a keyboard, a specific data value is sent when a key, or combination of keys, is pressed. This is usually kept in a temporary memory buffer that can be read by another program that is currently accepting input from keyboards (maybe a word processing program, maybe a browser window). A mouse sends two types of information -- a change in location and key presses. The key presses are handled similar to those of a keyboard. However, the change in location is done by the computer keeping track of the "location" of the mouse. When the computer starts up, the mouse is considered to be in a "default" (starting) location (often the upper left of your monitor). If you move your mouse to the right, it keeps track of how far to the right it has moved. Note that it isn't usually a one-to-one movement or your mouse pad would have to be as large as your monitor. Also, if you pick up the mouse and move it, it is as if it never moved.

Printers and monitors are typical output peripherals. Printers are fairly straightforward (although the actual data may not be such) -- your output may say "give me a new page", "print the letter 'a'", "go to the next line", and so forth but it basically is given a set of commands in sequence. A monitor, currently, is more complicated because of the idea of active windows and locations. So, in a modern operating system, the computer not only needs to keep track of where the mouse (as reflected by the **cursor** displayed on the monitor) but what program is making use of input while the cursor is at that location and what menu or button must be activated if there is further input (keyboard or mouse click) while at a specific location within that program's active space (or **window**). The input devices are actually what are causing changes about what you see -- the monitor just reflects the effects of that input.

13.   What Makes High Tech "high"? – 9/5/2009

People, and the media, talk a lot about "high tech"? But what *is* high tech? Warren Buffett says that he does not invest in high tech -- not because he thinks that it is bad but because he feels that it is wisest to invest in companies that one understands. This allows a good understanding of the business and the market and the managerial capability to run the company.

High tech exists because of all of the levels of technology upon which it relies. Thus, it is at the **apex** (or top) of a very large pyramid. This is reflected by our education and what we are expected to be able to do at different periods of our education.

At the bottom of the pyramid are "simple machines" and, as listed in Susan Kristoff's "Introduction to Simple Machines", scientists in the Renaissance period listed six devices for simple machines. These are the lever, the wheel and axle, the pulley, the inclined plane, the wedge, and the screw. In order to create, or make use of, simple machines, a variety of "simple tools" are required -- such as hammers, screwdrivers, wrenches, pliers, saws, and shovels. To this list, I would add cutting instruments such as knives and chisels.

Of course, when you read through lists such as this, you will find yourself saying -- but what about scissors, and bolts, and nuts, and ...? Just because the Renaissance scientists made a specific list of six items doesn't mean that everyone will feel that is THE correct list. For example, I

31

could argue that the wedge and the inclined plane are variations on the same thing. I would actually be tempted to classify into **categories** of mechanical movement such as lifting, pushing, pulling, and rotating.

It really doesn't matter. The fact is that there are some basic machines and tools that exist at the bottom of the pyramid. By using such basic devices, we can produce "low tech" -- a swing set, a door, a teeter-totter (for those that remember such), pogo sticks, and so forth. These can all be easily built using basic devices and tools. Of course, the degree of "finish" for something will depend on the quality of the tools and the experience of the craftsperson. I can create hinges and a door but you probably wouldn't want it on the front of your house.

We now reach into the "middle technology" category. Note that things are not REALLY so nicely divided as something may easily be broken into parts that are of various levels of needs. When we get to middle technology, we get to phonographs (for those who remember such), steam engines, paddlewheels, flour mills, ships, and so forth. These are items where you might appreciate the work which goes into them but aren't likely to cause you to gasp with astonishment. And many moderately experienced modern young adults can probably make such. They take knowledge, tools, and the ability to design and follow directions. I could probably make a horrible sounding, but functional, record player (and record). Neither Bose nor Boston Acoustics will lose sleep over that.

Now, we come to high tech. High tech requires the use of middle tech tools and machines to create their products. A CD player requires the ability to create lasers, and smoothly precise rotating movements, and a lot of semiconductor chips, and wiring and special metals and materials. I could NOT create a CD player without first creating a lot of things that I would need prior to being able to create the CD player. High tech requires middle tech which requires low tech. And that is the real story -- high tech builds upon other techs.

14.   Basics of How Email Works – 9/6/2009

I have noticed that some of my friends are not really aware of just how email works. This isn't really necessary -- as long as it works the way one "expects". It only becomes a problem when it does NOT work the way that it is expected.

Email starts out the same way as paper mail. It has an address and it has a letter (or **body**). It may even include some type of contents (**attachment**). The actual format of an email body is only important for a corporate systems administrator -- you'll probably never have to worry about it. The way attachments are carried may be of interest but, once again, it will work for you or it won't work for you. One popular method is **MIME** (Multipurpose Internet Mail Extensions) which allows mixing various types of files (photos, movies, music, text, etc.) within the same email file.

An address consists of a **user id** an "at" sign and a **mail server** name. A **server** is a computer which provides services to other computers (possibly in addition to providing services to someone directly using the computer). These services can be acting as remote data servers or it can be as a mail server. A mail server acts as a post office with each individual user id acting to identify an individual post office box. Popular mail servers include gmail.com, att.net, verizon.net, charter.net, yahoo.com, and aol.com. These people-readable, "friendly", names are actually translated into machine-friendly numbers (such as 103.56.113.114) by a system of other servers which provide **domain name services** (DNS). These machine addresses are used by the **Internet Protocol** (IP) network to route messages to the correct servers.

Most mail servers act as a "post office" for many different user ids. However, it is always possible for you to have a computer in your own home that acts directly as a mail server for your own **domain name**. When you send mail to someone, it is routed to the mail server for that address. The computer that acts as the mail server may be located anywhere -- Ohio, Paris, London, Florida, Washington state, or wherever. The mail will stay on that server until someone "picks it up" and then deletes it (or the mailbox capacity is exceeded). The important point is that, for most people, the email actually exists on some other computer that may be far away.

In order to pick up your email, you have two basic options. You can use a browser (or other program) to connect directly to your mail server and read/delete/send mail. Or you can use a mail program (such as Outlook, Thunderbird, Mac Mail, or AOL) to connect to the mail server and **download** the mail on the server to the computer upon which you are typing. At the point of downloading, there are still two copies of the email -- one on the mail server and one on your local computer. However, this is a waste of storage space, so most mail programs give you various options (under "options" or "preferences" usually) to delete the copy of the email on the mail server. This can be done immediately, in three days, in a week, or an arbitrary interval.

And now you have your email and can deliberate on a thoughtful response.

15.   Ready, Set, Stop – 2/19/2010

We've probably all seen a car in an intersection -- rear end over the crosswalk and front end partially blocking the outside cross traffic land -- and they just stay there, never getting a green light to go. Or perhaps they have stopped 30 feet behind the crosswalk and they're stuck (and you're stuck behind them). There really is no mystery -- they are beyond the range of the sensor in the road and the traffic light system doesn't know they exist.

Traffic light systems are ideal for computer programming assignments. The basic system is very simple but it can be increased in complexity to understand more and more possibilities of design. I have used such systems as examples in a couple of my books.

The simplest form of traffic light is a blinking four-way stop (red lights). Not much of an advantage over a four-way set of stop signs except more visible in the dark. The next version can make use of a simple mechanical timer and set of switches (rather like many mechanical pool pump timers). The timer can revolve at a fixed rate and close contacts with the appropriate lights in the traffic system. It probably proceeds like

Green Red
-t- Red
Yellow Red
Red Red
Red Green
Red -t-
Red Yellow
Red Red

and back to the beginning. The "-t-" indicates some type of delay -- the length of time that the green light stays on for that direction. Note the two times that both directions are Red. This is very important for safety reasons.

This simple mechanical system has a timer and a set of connections to lights. The timer is an input and the connections are outputs. Mechanical systems can be designed to allow for a beautifully complex set of conditions but the actual construction becomes more and more precise and difficult to mass manufacture. It is much cheaper, and easier, to start adding microprocessors and programs to handle more complex operations.

With programming, the inputs are often extended to a clock, a set of timers, and one or more sensors. These are all concerned with **events** that affect the output -- which may be extended to include walk lights in addition to traffic lights. The programming may start taking into account the day of the week, time of day, whether it is a holiday, how many cars are waiting in a lane, and many other options. Once again, it can get pretty complex -- but most of the complexity is hidden in the programming and, thus, mass manufacturing is still possible (even reduces the cost per unit when more are made).

So, the car is stuck because the sensor no longer can tell it's there and it cannot use that as an event to trigger a light change.

Why go into all this detail? Who cares? Well, I find it interesting in itself but it's also a good prelude to talking about embedded processors in cars and the relation between complexity, sufficient testing, and safety which is definitely in the news of late. See the next blog.

16. Where Did They Go? – 4/17/2010

Well, I decided to be inspired by one of the Blogs I follow - - "The Retirement Bubble" and accept the fact that I just am not going to be a daily blogger. So, I headed to my blog and "ZAP", my last blog was no longer there. In fact, my stat counter for visits (which I watch not go up very quickly) went backwards from 184 to 121.

I'm sure that things of this nature happen to you, also. What can the reason be?

Well, first, of course, there is the jello-like consistency of memory. I could have just imagined that I posted a blog entry last month. People look at memory as the chronicle of the past but it just doesn't really work that way. If you think about doing something enough times, with enough detail, it will blur the boundary between "memory" and "dream". Given an amount of elapsed time, that boundary may easily disappear. I know people who have very fixed "world views" and you can tell them "yes" to a question and, because they just "knew" you were going to say "no", they will HEAR "no" and remember "no". This is a bit more severe of a split between "memory" and "dream". The bottom line is -- one cannot really rely on memory

But, do I think that is what happened? No. Of course it could be vanity -- "other people may not remember correctly but I certainly do". No, the main reason I don't think that was the case is because of the stat number. Everyone has certain areas where their memory is well exercised and more reliable. My stepdaughter can remember what someone wore for a given date within the past few months and practically forever about what SHE was wearing. Other things aren't so important to her and she just doesn't remember. For me, it is numbers. I can visualize that "184" in the stats area and I'm pretty sure it isn't a false memory.

There are some aspects to irregular posting that are certainly suspect. For example, some of the formatting aspects of this blog seem to be strange to me -- but, for that, I will just blame my memory and not doing this blog often enough.

Assuming that I'm not crazy and my memory is not totally faulty, where did the blog go? Two possible avenues seem to rise to the surface. One, a system crashed and the disk got backed up (which, for Google, does seem a bit scary to think that backups are unreliable). Two, I didn't do something correctly to commit the blog into permanent status.

Actually, there is a third possibility but paranoia just isn't my thing. "Someone" could have removed it. Since the blog was concerned with firmware and quality control (and Toyota) I guess that there is a little weight to that but I don't really believe it.

At any rate, my last blog vanished and I'll have to think about it a bit to start over the chain. The last blog was on embedded software (firmware) and quality control. The next was going to be about embedded software on cars, unit testing and system testing -- and the difficulties of fully system testing real-time software for interconnecting modules (groups of software). And the next was going to be quality control and testing in general.

However, since the first of the series has vanished I guess I'll think it all out again.

Google, if you're listening -- maybe YOU can find out what happened to my last blog.

17.   Hackers are Positive People – 9/29/2010

I was looking around at the movies coming up and looked into "The Social Network" which led me to Wikipedia for Mark Zuckerberg.

Once upon a time, I wrote a book about how computers work -- meant for the everyday person and trying to explain/show how the various parts of computers worked together. Although unpublished (and probably unnecessary anymore), one of the chapters dealt with hackers. According to Wikipedia, Zuckerberg said that hackers are people who aren't afraid to break something in order to make it better.

Personally, I would go a bit further. I would say that hackers are people who are curious about how software works, aren't afraid to break it, and who want to make it better (in the past, this often meant more efficient and faster, nowadays it more often means more features). At any rate, it is a positive purpose -- Bill Gates Jr., Wozniak (also a hardware tinkerer), and others are all hackers. I am also a hacker and proud of it.

So, why does the word "hacker" have such a negative connotation? Leave it to the sound bite. It sounds good and the media, rather than going into more detail as to what was actually done (which might bore a majority of their viewers/readers) just lumped all people working on software as hackers. This includes a sub-group of hackers which does NOT have such benevolent motives -- the "crackers". Crackers work to exploit the inherent weaknesses of software structure to allow misuse.

Crackers are a pain in the rear -- they do things for their own egos and pocketbooks and make the rest of us have to deal with their behavior. Hackers have made the fusionfalls (a current favorite of my kids), the facebooks, the Linuxes, and the spreadsheets possible (among thousands of other software programs). Hackers have found a home in the open source community but, for the most part, remain in individual obscurity.

18.   The Tree of Data Connectivity – 6/8/2011

The goal of data connectivity is to allow people to communicate with everyone else. The form of the connection may be voice, text, video, files, or some other type of data. However, the ideal is that it can transfer between any two (or more) people.

So, how does that happen? Well, the actual form of the data is in files that have some type of format. These formats, or organized forms, are standardized so that it can be created and transmitted, and then received and used. Examples of these are Adobe PDF files, Audio MP3 files, and Video MPEG-4 files. Voice is usually digitally encoded using one of the International Telecommunication Union (ITU) standardized formats such as G.711 or G.729.

Once a file has been created, it is transferred to the desired recipient. The current dominant types of transferring may be classified into wireless and landline. Wireless includes cellular data (and voice), Bluetooth, and Wi-Fi. Landline (using a physical medium for transfer such as a cable or fiberoptic line) makes use of many different protocols.

Wireless is the current popular option for the "end user" -- the person who is making use of the device that allows them to transfer data (remember that voice is just one form of data). However, wireless has significant restrictions. These restrictions are distance and/or speed. Companies are now marketing devices called "femtocells" (the actual brand and name of the device may not include this name) which are placed in homes or businesses to allow faster transmission by reducing the distance between the cell phone or Wi-Fi device and a landline.

Yes, a landline. Because of transmission requirements (which go beyond the scope of this current blog), it is just impossible to handle all of the data needs without going to a landline. A landline is self-contained for bandwidth (data capacity) while wireless must compete with all other wireless activity. So, landlines will (in my limited ability to read a crystal ball) always be needed and they form the trunks and branches of the tree of data connectivity with the wireless devices now being used as the leaves.

19.   Old Memory – 7/15/2011

Yes, I walked a mile through the snow to school. But, that really wasn't such a big deal (still isn't). I lived in a small town where walking was quite safe and there were sidewalks in front of almost every house.

But, in terms of computers, I started programming when the main input media was punched cards (I saved a lot of them for shopping lists for a number of years). When I started college, I worked on a "personal" computer that used punched tape along with toggle switches on the main processing unit. The main storage devices for the larger computers (IBM 360 at first, migrating to IBM 370s before I left) were huge disk drives.

However, memory (as mentioned before) can be categorized into temporary working memory and long-term storage memory. Working memory at that time was **core** memory -- little magnets that looked like donuts and linked together with copper wires. We still call it core memory after the memory of those days. Most working memory of today is now located within DIMMs (Dual In-line Memory Modules -- see Wikipedia) or, for older systems, SIMMs.

The memory modules have great advantages over old core -- speed, size, and capacity. They also generate much less heat which is both an energy savings as well as a design improvement.

Storage memory is another category which has moved from technology to technology. The first was paper (well, the VERY first was probably clay tablets or chiseled stone).

For people, writing or drawing stored the data/information and reading brought it back. It's kind of funny, but efforts have been quite intense over the past twenty years to allow computers to do the same thing that humans have done -- to be able to directly make use of printed text and images.

For early computer systems, it was not possible for the computers to directly use text or drawings. They needed a way to detect a contrast between spaces. This usually meant holes. The holes allowed light, or a mechanical probe, to move through the paper. The areas without holes blocked the light or probe. In this way, the computer could read the "bits" (present/non-present, on/off, 1/0) and save it.

Next came magnetic methods. These were primarily on discs and tapes. The technology of disk drive design has developed enormously over the past 30 years such that a portable disk can hold the same data that a room of luggable, replaceable disks did way back when.

Currently, we are moving farther and farther along to making working memory cost-effective to use as storage memory -- which will lead to the next post on "New and future Memory".

20. New and Future Memory – 5/14/2012

I've got a few more ideas to talk about, but I thought it best to come through with what I said I would do next.

There are two trends going on with memory (for computers -- there is also some fascinating research going on about human memory) nowadays. The first trend is putting it elsewhere (in the "clouds") and the second trend is to eliminate the mechanical aspects of data storage and access.

I will push off the discussion of clouds to the next blog. We'll concentrate on the second trend.

There are a lot of excellent disk drives at the moment. The manufacturers have increased storage capacity, decreased the time to get to (access) the data, and greatly improved reliability.

Most improvements on disk drives have been associated with data density -- how many bits can be packed into the smallest area. The data density helps both storage capacity and transfer rates (the amount of time that is needed to move data from the storage device to working memory (or vice versa)). Blu-Ray disks work with a higher-frequency laser than do DVDs and DVDs use a higher-frequency laser than Compact Disks (CDs). The higher frequency means that the data density can be higher. Thus, Blu-Ray disks can hold more data than DVDs and DVDs more than CDs.

Further improvements are being made on materials, optics (the part that actually reads optical disks such as Blu-

Ray), Wikipedia is a great source for more on specific formats and improvements.

What do these disks have in common? They have to move. In order to read (or write) the data, the reader ("sensor") must be over the datum. Usually, this means spinning the disk while the reader stays in the same place. Some magnetic hard disks have speeds exceeding 7800 revolutions per minute. However, movement means something to move it with and mechanical devices just will not work forever no matter how great the quality and design.

We now have many different electronic items -- phones, cameras, tablets, toys, and so forth that make use of non-moving memory. There are a lot of different categories for this, so let's just call them "flash" memory. In this case, there are still lots of data to access -- but the access method is built into the design of the memory device. Let's take a game cartridge as a simple example. The cartridge will contain data which can be addressed. It also has leads (usually copper) that connect to the game player. The game player makes use of these leads to address, and transfer, the data. No physical movement (except for connecting the cartridge to the game player) is required. Another common example is a "memory card" which is inserted into a camera. Some printers allow photos to be directly printed from that memory card (taking the card out of the camera and inserting it into the printer).

Direct access memory devices are (currently) more expensive that disk drives -- but the cost continues to decrease as they become more popular and it is my opinion that they will take over for local storage eventually. Personally, I am still hoping for holographic cube storage as was seen in Star Trek.

Let it be so.

21.   Here Come the Clouds –
5/28/2012

Once upon a time, I had a Cathode Ray Tube (CRT --
similar to old televisions) terminal that was connected to a
MODEM (MODulator DEModulator -- a box that converted
digital signals to/from analog signals on a phone line) which
connected to a "mainframe" (very large -- for those days --
computer). My terminal, an ADM 3A, was a monocolor
screen with a keyboard. No local processor, no local storage.
When I logged into my account on the mainframe, I had my
own set of files and access to any applications that had been
installed on the main computer.

Now, with the computing clouds, I can have a computer
(possibly without a disk drive/local storage) of, perhaps,
limited computing power connected to the Internet which
gives access to one or more main computers and multiple
storage areas. I can connect via different devices and from
different locations and get access to my own set of files and
make use of various applications installed on those
devices/computers.

It sounds very similar between 1981 and 2012 doesn't it?
It certainly does to me. What are the differences that exist
and what makes those differences?

The first is access speed. In 1981, connections were slow.
A person could type in a set of commands (no Graphical
User Interface (GUI)) and expect to receive back sets of
words or numbers -- possibly some crude pictures made up
of typeable characters. The terminal would allow some
movement of the "cursor" (think of the marker from the
mouse) after receiving special characters that would be

interpreted specially. But, basically, it was for sending and receiving text.

In 2012, expectations of connection speed are FAST or FASTER. This means that the user can use a GUI and can receive back all forms of data including video, music, and multiple windows of information.

The second difference is primarily on the "mainframe" side. Via the Internet, the user has access to many different computers, different environments, and a multitude of applications. Thus, the user can treat the "cloud" environment as their own individual computing setup. Plus, since the computer that one uses usually has its own memory (even without a disk), work can be divided between the local computer/PC and the cloud devices.

From an outside point of view, 1981 and 2012 seem rather similar. The effect of access speed and the Internet's capability of hiding where and what is happening creates a very different experience.

The next blog will go into greater details on the advantages (and disadvantages) of the cloud environment.

22.   So, What is a Cloud? – 6/1/2012

Clouds have been around for a long time. No, I'm not talking about the groups of water droplets that sometimes are between us and the sky. The cloud has been the nickname for the general network for a long time. When a picture was drawn of two people talking together over the phone system, the picture usually had the originator (call this person 'A') talking on a phone which had a line to a cloud-shaped symbol which then had another line leading out of it to the recipient (call this person 'B') of the call. When a physical connection exists between A and B, it is called a "circuit-switched" line.

Over the years, what has actually been within that cloud has changed. Long, long ago, the contents of the cloud were a series of connected wires such that, physically, there was a single wire leading from A to B. In order to achieve this connection, various people ("operators") would use a small section of wire (called a "patch cord") to connect lengths of wire together. So, your local operator (which had ALL the local phone wires leading into the office) would connect your wire to a wire leading to a long-distance operator, who would then connect to the destination region, who would connect to a destination city who would connect to a local phone company who would then connect to B's line and then put a "ringing signal" on the line to tell B that they had a call.

The next iteration of content in the cloud was to replace part (then all) of the human operators with mechanical analog (no bits and bytes) switches. One switch type, called a "cross-bar" was an important development that allowed

49

this progression to change. There was still, by the time a call was completed, a single physical connection from A to B.

The change from analog to digital allowed "breakage" of the physical connection. While there was still, after the call was completed, a physical connection, the form of the signal now changed from section to section of the connection. This usually meant a parallel line that contained "signal" information. The signal information includes such things as to whom the call has been placed, who made the call, when it occurred and (for billing purposes, in particular) how long the call was active.

Packet-switching broke the physical connection. Packet switching includes the address information (telephone number, etc) with the data (voice, video, music, ...). Since the address was included along with the data, it could be sent anywhere -- it could even be stored temporarily if a connection was unavailable. Finally, the Internet Protocol (IP) started to take over this type of combined address/data format.

However, when you make a call (or, access a computer or network service or whatever), you don't really know what is happening in the network -- and that is why it is still envisioned as a cloud. And, you don't really CARE how it gets from A to B as long as it gets there. It is likely to be a mixture of technologies and it just isn't important to A or B -- but it is vitally important to the providers of the network.

Modern "cloud" services rely on a packet-switched Internet Protocol network to allow access, storage, transfer, and interpretation of data. The next blog will talk about some of those specific services.

23. What Does the Cloud Provide?
– 6/8/2012

OK. We have seen that the cloud is a nickname for the potentially changing and somewhat mysterious connections that allow equipment (and people) to talk and send data to each other. The modern cloud will usually make direct use of the Internet Protocol (IP) network -- although that is not mandatory. The advantage to using the IP network is that each request can be routed to a different location.

In the old cloud, you basically had a direct connection (often called "point-to-point") between two pieces of equipment (possibly phones). With an IP network, since each message contains the address of the originator and the address of the destination, the resulting connections are "many-to-many". Your local equipment probably has a single IP address but the IP address is used in conjunction with another piece of information called the "port". The IP address identifies the physical device that is receiving and transmitting data and the port is used for routing the data to the right application or task.

What does this mean in real life? Let's say that you have a word processing application open and you also want to listen to music while you are typing on the document. The word processing app might be using a combination address of "53.13.18.01:2022" where the part before the colon (":") is the IP address and the part after is the port number. The music application makes use of "53.13.18.01:1954" (these are arbitrary numbers). Since the two apps are making use of two distinct ports, the data can be routed appropriately.

On the other end, the word processing app might be connected to "103.44.17.34:1113" and the music app is getting the music (data) from "87.19.33.92:1954". Note that the music app and its data are using the same port number -- it's not required but it does simplify some of the interactions. We can see from the addresses that we have two applications on a single physical device connected to two separate data providers which are likely on separate physical devices.

This is the power of the cloud -- the physical and logical separation of the data from the applications making use of the data. The data storage might be of music, documents, spreadsheets, ebooks, or whatever else you can imagine.

Next, what about applications in the cloud (sometimes referred to as "Software as a Service" or SaaS)? Well, actually, the data providers are applications and are interpreting the data coming from the "local" application in order to retrieve and route data appropriately. SaaS moves most of the processing of the data to the remote server. It isn't actually in the cloud but, from the point-of-view of the local user, it may still be located anywhere and, thus, part of the cloud from one endpoint's point-of-view.

Finally, the cloud can provide alternative paths and destinations. This can provide data transparent backup. Let's say that you have your endpoint making use of data stored at location C. Unknown to the user, C is constantly backing up ("mirroring") the data at location D. If the physical device hosting C goes down (is now unavailable) then the local app can be routed to D without the user even knowing anything has gone wrong.

The cloud provides many services and will provide even more in the future. However, with this complexity comes different types of vulnerability. I will address that in the next blog.

24.　　The Dangers of Hi-Tech –
6/15/2012

I am **NOT** a Luddite. I enjoy technology and, even more, I love the thinking processes involved in creating something new and tweaking (called "hacking" in programming) existing things to make them perform even better.

However, that does not mean that I am not always aware of the dangers of advanced technology. Sure, there is the Frankenstein aspects -- designing "smart" robots that take over the world. I'm not that awfully fearful of that -- although it certainly could happen.

The Luddites were more fearful of how technology affects society and how new things cause the old things to be denigrated. What happens to the horses when a steam engine can pull the plow? Each change in technology affects the society and people involved with the old technology. That is still true today and, perhaps, even more true as the pace of change continues to increase. Still, that has always happened and it always causes chaos. The first people who made use of iron weapons were ascendent over those who still used stone weapons and the people who used bronze weapons were ascendent over those using iron and so forth. This has not changed -- only the rate of change where now the change can occur many times in one's lifetime rather than over a period of several generations.

My primary fears are the pyramid effect of Hi-Tech.

If I worked carefully over a period of a few weeks, I could make a gramophone. (This is one of those old "record players" with a big copper funnel over the needle to make the noise louder.). With more time and access to more tools, I could probably make a "record player" that connects to a home-built electric speaker. It wouldn't be of great quality but, with practice, I could make pretty good ones. But could I make a ceramic cartridge magnetic needle casing with a full powered amplifier and multiple speakers? Yes -- but only by using a lot of other tools that are a per-requisite to make it. Over a lifetime, I might be able to create the entire set of tools and then use those tools for the final product. A lifetime wouldn't be enough to create a CD player.

My father would have been even better qualified to build something than I am -- among his many jobs, he was a machinist and an auto mechanic. He could have built a working carburetor from chunks of metal. However, even he could not have mined the ore, smelted the ore, refined the metal and created the metal chunks that he needed.

Every "hi-tech" product relies on components that, in themselves, are "hi-tech" and requires specialized tools to build. On and on down the line.

I have a "landline" phone in addition to my cell phone. I have no intentions of giving it up (although economics may eliminate them as an option someday in the future). Why? First, the sound quality of a dedicated circuit-switched line is better than anything you can currently find in cell phones and probably better than you ever WILL find in cell phones. But, more importantly to me, a basic landline phone is powered by the line that leads to the phone company. They have banks of batteries to supply the very low voltage current needed to power the phones. If the electricity goes out -- I still have a working phone! Even there, most people have landline phones that are connected to local electricity -- and those won't work without the power. They COULD be designed to make use of the line power but most aren't.

So, what is the danger? The danger is that a disruption in one vital element of a product will eliminate the feasibility of the product. Many things are dependent on oil products -- run out of (or be separated from the access to) gas/oil/diesel and much of society's products will stop working. Have an electromagnetic pulse take place in New

54

York City and much of our economic records would disappear -- even worse if backup sites are attacked/broken.

So, when you pick up your smart phone, think a bit about what it really took to make it and be able to use. There're a lot of industries, professions, and people involved in that one product. Then turn on a light and do the same type of thinking -- still a lot of factors involved in the use of that light switch. If you have a gas cooktop you are reliant on a steady gas supply but if you have an electric cooktop you have a different set of dependencies.

25.   Updating an Electronic Device - 7/3/2012

A few months ago, I talked about computer memory and its various types. One of the important types is Read Only Memory (ROM). This contains the basic instructions (including the instruction that is executed first when power is applied to the device) to bring up all of the supporting programs (Operating System, etc.) that allow you to do what you want to do.

What happens if you have a device (such as a cellular phone or a game system) that is "updatable"? This device also has ROM but, somehow, the basic system that is in ROM is still able to be changed.

This is possible because there are different types of ROM. In particular, there is a type called "Electrically Erasable Programmable Read-Only Memory" (EEPROM). Like general ROM, this memory is non-volatile -- it will retain its contents even when there is no power. However, by applying a higher-than-normal power through the device, the contents can be erased and then new contents can be written. Thus, a device which is meant to be upgradable can split its base program memory into two parts -- one in ROM which still contains the initial program upon powering up the device and one in EEPROM which should not normally be altered either during use or when powered down.

The program in ROM is enhanced to include the program(s) that allow updating the EEPROM. Then, when an update is desired, it stores the new system program to be written to the EEPROM in some type of RAM, erases the original contents of the EEPROM and then copies over the new system program into the EEPROM.

There are variations on this, of course. When a device is updated via a host computer, there is the option to back up the data and current contents before doing the update. This reduces the danger involved in case power is interrupted, or an error occurs, before the update can complete. The time when an update is in progress is still a critical period of time and, if at all possible, the device should remain powered up until updating is complete.

Once the update is complete, the device should operate the same as before (with whatever improvements exist within the new system program).

26.   The Science and Economics of Nutrition, Part 1 -8/25/2012

In some ways, this is not about what most people would call technology. However, the science behind nutrition is something that can be useful to all of us. There are a lot of people who do not understand just what is involved with "good nutrition" and they rely on aids, such as food pyramids, to attempt to create a good diet. However, understanding allows making better choices and knowing why we make those choices.

A healthy diet is composed of two factors -- calories and nutrition. On average, an adult needs approximately 15 calories per pound to maintain their weight. There's a lot of variation on this. Athletes and pregnant or lactating women need more. Sedentary people need less. But let's go with the 15 calories (actually kilocalories -- but most people just call them calories) per pound. This means that a 120 pound person (or someone who wants to get to 120 pounds) needs 1800 calories per day.

Calories are a measurement of the energy from food and, once again, it is simplified in presentation. A gram of fat is about 9 calories and a gram of protein or carbohydrates is about 4 calories. In general, it doesn't matter what kind of fat it is -- it will give you the same amount of calories. So, olive oil may contain "better" (from a nutritional point of view) calories but it will still be the same amount as that from lard. Proteins, once again, are all about the same for calories but the ability to be "burned" (metabolized) varies depending on the mixture of other foods with the protein.

At any rate, fats are the most calorie-dense foods at 9 calories per gram. This means that an average 120-pound person could get their daily calories from drinking 18 tablespoons (1 1/8 cups) of oil. They would also get severe diarrhea and eventually die of other causes -- but they would have enough calories. Prices of food vary around the world but, in the U.S., you can get canola oil at about $10 per gallon. There are 16 cups per gallon, so this amount of oil would cost about 70 cents.

At the high end of the scale -- I just don't know what that would be -- probably some rare gourmet low-calorie item. Let's just say that you can probably spend more than a $1000 for your 1800 calories. This is talking about actual base food costs -- eating at a restaurant would certainly increase your costs.

OK. We see that you can spend from 70 cents up to thousands of dollars to satisfy your caloric needs. But, we said at the beginning that a healthy diet is composed of calories AND nutrition. What is nutrition? These are the various components that your body needs to be as healthy as possible. It includes a proper balance of fats (and the right kinds of fats), protein (and the right kinds of protein), and carbohydrates (and the right kinds of carbohydrates). It also requires minerals, vitamins, and dietary fiber.

In the next post, I will start talking about the nutrition aspect of a healthy diet along with the economic impact of choices.

27.   The Science and Economics of Nutrion, Part 2 – 9/18/2012

"You are what you eat". This is basically true although, if you eat an apple, you are safe about not having little bits of apple roaming about your body. A good way to look at it is that what you eat provides building blocks for your body to use -- both keeping the body at operating temperature as well as to replace old cells and build new ones. The system of doing this is called metabolism and it is not really well understood although there is a new theory (and a new diet) on a regular basis.

Most of the nutritional studies are of an experimental kind. This doesn't usually mean a room full of lab rats (although such may be involved). It means that the people examine what goes in and then monitor effects that can be measured. The "in-between" mechanisms are still rather mysterious -- but there are some fascinating studies that have been published in the last few years about the symbiotic roles of bacteria, viruses, and other "critters" that exist on our skin, within our digestive track and even within our cells and organs. The mitochondria, that exist within all of our cells and are an integral part of metabolism, probably originated as one of the first symbiotic organisms.

One thing to be very careful about when considering nutrition is that the food brought in is broken down and then reused (building blocks, remember). If a person eats cholesterol (a fat with a bad reputation), it does NOT get directly rerouted to the arteries. You can eat all the protein you want and, without doing other things, it will not automatically give you huge muscles.

We mentioned in the previous blog that nutrition is involved with the availability of various components in the diet -- including fats, proteins, carbohydrates, minerals, vitamins, and dietary fiber (actually, soluble dietary fiber). This blog will focus on fat which probably has the worst Public Relations system of all.

It would probably help the reputation of fat if there were two words for different situations. Call the fat that you consume "oils" and the fat that you possibly have in excess around your middle as "reserve materials". Fat is very important in our diets and our bodies cannot produce healthy cells without it.

Oils are organic solvents. This means that organic materials tend to be able to dissolve in, and disperse within, fats or oils. This is why fat added to cooking aids the flavor significantly. Spices and other materials that cause us to say "yummy" are dispersed within the fat. This same property is necessary within our metabolism -- the fats provide a "kettle" that can be used to concoct all kinds of cells. One important class of cells that requires fat is our nervous system. A lack of sufficient fat in the diet can cause nerve damage and is why pediatricians emphasize to NOT give fat-free milk to infants -- they need the fat to help produce healthy brains and a healthy nervous system and do not have other sources for the fat.

There is rarely a consensus (everyone agreeing) within the nutrition field but, in general, it is recommended that about 25% of the calories we consume come from fats. Since fats have a bit more than twice the calories per unit as proteins and carbohydrates, this means about 12% of our food, by weight, should be fats. Excess fat, by itself, will not directly lead to greater "reserve materials" around the waist.

Our bodies are very wise and they know what we need. We are designed as omnivores -- so the easiest way that our bodies can grab all the needed building blocks from our diet is to eat a good assortment of foods, including meat, vegetables, and grains. Vegetarians, and vegans, can also have a very healthy diet (some say healthier) but the body alone cannot grab all the needed building blocks without some careful planning of our diets -- the building blocks that come from meat (in particular, a balance of "essential" amino acids) must be replaced by equivalent building blocks from other foods. If we get all we need, the excess will be removed from our bodies. However, if we get an excess of calories (and have the base foundation nutrition that we need), the body probably WILL make use of some of the building blocks to put down a layer of those "reserve materials".

The area of greatest controversy and the fastest shifting arena within nutrition is the battle of "good fats" and "bad fats". Once upon a time, people were encouraged to eat margarine and avoid all the "bad fats" of butter which comes from animals ("moo"). Currently, butter is considered to be better than margarine because, in the process of making margarine solid, it creates certain "bad fats" known as Trans-fats. Butter also has trans-fats but less than most solid margarine and more "natural" since it was created by an animal. There are saturated fats, unsaturated fats, polyunsaturated fats, monosaturated fats, Omega-3 fats, and so forth. If you want to know more about these, there are lots of good books about various fats out there. Right now, the general idea is that the less saturated that the fat is the better it is for you.

So, what is the best fat for you? This is where economics rears its head again. A person can get oils from fruits (olives), vegetables (sunflowers), legumes (peanuts), grains (rapeseed -- canola), and animals (lard as well as Omega-3 oils). Many recipes and references just split these oils into vegetable oils and animal fats.

Canola oil is produced heavily and is, therefore, one of the cheapest oils but there is some controversy over its genetic history. Omega-3 oils are in much smaller supply and are, therefore, one of the more expensive oils. Since this area is most in flux, I certainly won't try to recommend a specific oil but, if you look into the pantry of people, the type(s) of oils you find will certainly reflect income level to a considerable degree -- and some say the more expensive oils are "better".

28. The Science and Economics of Nutrition, Part 3 – 10/19/2012

When it comes to the area of proteins, the description of food as building blocks is even more directly true than ever. Although most of us think of proteins in terms of muscle and (possibly) hair/fingernails, proteins are an important part of our entire body -- from "scaffolding" for cell walls to acting as catalysts for general digestion.

However, protein eaten does not directly translate to protein built in the body. The body's digestive processes break down proteins into building blocks called amino acids. It then uses these amino acids, as determined by specific gene sequences, to create the proteins the body needs. There are 20 or 21 (depending on how you classify them) amino acids used within the human body. Some of these can be created by the body from general food (that is, not amino acids). Others can be created from other amino acids (changed from one to another). Some, however, must be eaten and these are the "essential" amino acids.

Essential amino acids must be part of a regular diet -- but that doesn't mean they all have to be eaten each day. The body just must have a reserve of them in the "storehouse".

Meat has the advantage in that the animal has already gathered up, or created, the various amino acids needed. However, it is fully possible to get all the needed amino acids from a varied vegetarian diet. The mixes needed to have a complete set are called "complementary" foods. Beans and rice work as "complementary" foods. Lentils and barley are a great combination. There are many others.

As mentioned, meats give "complete" sets of proteins -- whether it be beef, poultry, fish, pork or some other meat from animals. Beef often gets a bad name as a meat. This is not because of the protein but, rather, from the saturated fat that is often mixed in with the protein. Depending on the particular cut of meat and the way that it is prepared, it is possible to have very lean, healthy, beef as a part of a meal. In fact, if you prepare poultry in a high-fat manner (think fried chicken), chicken can have higher levels of fat than beef. Fish is often separated from other types of meat because the fat which it contains is normally UNsaturated (including "Omega-3" (linoleic) acids). Thus, fish protein is lean and the oils that come with it are recommended types of oils.

In the area of economics, meat is an expensive choice. This is largely because it is higher on the "food chain". In the U.S., at the grocery store, chicken can average $3/pound, turkey $1.50/pound, fresh fish may cost $7/pound (depending on geographic location). It is difficult to spend less than $1.50/pound for meat but it is possible to spend more than $50/pound for particular cuts of meat or specially prepared meat or varieties of animal meat. On the other hand, a combination of lentils and barley (enough for a single person) may only cost 30 cents ($0.30). If you choose the vegetarian route then you have three advantages -- cost, a light footprint on the earth (less of earth's resources used) and an automatic advantage in not getting "unhealthy" fats (allowing you to choose what fats to incorporate within your meals). The disadvantage is that you MUST vary your diet deliberately in order to have what your body needs.

One additional disadvantage of animal meats should be mentioned. This is the fact that most meats available at grocery stores are "factory meats". The animals have been treated as raw materials to produce meat. Not only are these methods not kind to the animals but the process requires antibiotics and hormones to keep the animals alive long enough to harvest. The factory meat process is probably the largest reason for waves of recalls of contaminated meat. (Note, however, that vegetable products are not immune to this -- the recent recall of peanut products.). While you can get non-factory meat, it will be more expensive and it still

66

requires a lot of food (particularly grain products) to produce.

I love a good hamburger -- and bacon must truly be set into a class of its own. However, I also love a good lentil and quinoa salad. We are designed to be omnivores and, as long as we are aware of what we eat, we can get good protein into our diet in many different ways.

29.   The Science and Economics of Nutrition, Part 4 – 11/6/2012

Carbohydrates are an important part of a general diet because they provide fuel for the body. They are not directly used as building blocks but provide energy for use of fats and proteins and incorporating minerals and vitamins into our bodies. They consist of only Carbon, Oxygen, and Hydrogen atoms -- thus, the name "carbohydrates" although they are not chemically considered to be "hydrates".

When I was double-checking my sources for this article, I found that I had incorrect ideas about alcohol. Alcohol (or, more specifically, ethanol -- drinking alcohol)
is **NOT** considered to be a carbohydrate in spite of having only carbon, oxygen, and hydrogen atoms. However, just as ethanol can be used to fuel machines, our bodies can make use of it as an energy source. Alcohol provides about 7 calories per gram (almost as caloric as fat). Alcohol is not considered to be nutritional and burning alcohol is hard on the body (the liver in particular) and should be used sparingly. A tablespoon of pure alcohol would be about 85 calories -- an 8 ounce glass of wine about 190 calories.

Carbohydrates are largely the same as saccharides. Saccharides are grouped into monosaccharides, disaccharides, oligosaccharides, and polysaccharides. The first two are usually referred to as "sugars" while the other ones have various names including "starches". Since carbohydrates do not provide any "building block" materials the amount can actually be fairly low with calories provided by fats and proteins -- but this is not really recommended. The range of percentage of calories provided by carbohydrates in the diet is suggested to be in the 45% to 65% region with simple "sugars" limited to around 10%.

It is difficult to discuss carbohydrates without bringing dietary fiber into the discussion. Dietary fiber is looked at as being in two categories -- soluble and insoluble fiber; "good carbs" are talked about versus "bad carbs". It really isn't a matter of the carbohydrates -- it is how they are utilized within the body. The metabolism of carbohydrates (and fats) is regulated by insulin and, thus, eating "simple" carbohydrates will cause a rapid rise of insulin in the body which is hard on the body and related to metabolic problems like diabetes. Soluble fiber works with carbohydrates to allow the digestion to be greatly slowed and extended (look at it as "time released" carbohydrates) and this allows the carbohydrates to be better utilized by the body. A system called the Glycemic Index is a good method of determining the gentleness of different carbohydrates in the diet. Note that insoluble fiber is of use, also, as it provides "roughage" to allow the muscles of the digestive and excremental tracts to be more effective.

One aspect of the Glycemic Index is that it is an isolative, or simplified, look at a single food source. It is possible to still have a gentle diet with simple carbohydrates if it is eaten WITH other foods that can supply the soluble fiber. Thus, rice is not really easy on the body -- but eating rice with high-fiber foods such as beans, or seaweed, or vegetables means that the entire meal is well balanced. This is why many diets that rely on rice are healthy -- it is because they are in combination with other foods which supply needed soluble fiber.

The economics of carbohydrates come into play because simple carbohydrates are inexpensive and, thus, are easily incorporated into unhealthy diets. More balanced carbohydrate sources, such as oats, barley, and beans are also fairly inexpensive but it requires more time to work with them. In the area of applied nutrition, time does equal money. Thus, many fast processed foods have a high percentage of simple carbohydrates and fats. In part six of this series, we will discuss how the choices for food can be made, could be made, and are normally made. (In part five,

70

we will finish up the nutritive discussion with vitamins and minerals.)

In summary, carbohydrates provided needed energy for the body to be active and build cells. In order to be used in a manner that is easier on the body, the entire meal must be planned and examined.

30.   The Science and Economics of Nutrition, Part 5 - 11/30/2012

In the past few posts, I have talked about the three caloric supports of nutrition -- fats, proteins, and carbohydrates. Carbohydrates only supply energy but fats and proteins also supply necessary "building blocks" for the body to build and repair itself. However, these are not sufficient -- it is also necessary to have other components which we usually call vitamins and minerals.

Vitamins are organic compounds that cannot be synthesized by the human body. Thus, some items (such as Vitamin C) are needed vitamin for humans but not considered such for other animals. Vitamins primarily act as catalysts for building processes within the body. This means that they help the body produce needed cells and substances but are not directly incorporated into the body.

Minerals are often thought of as the "basic elements" such as Calcium, Iron, Copper, and so forth. However, most minerals in the diet are actually combinations of elements -- often with Oxygen but possibly a combination of various elements such as Calcium and Carbon. Since Oxygen is such a high percentage (about 46.6% by weight) of the earth's crust, it has to be expected that Oxygen will be incorporated in many minerals.

Minerals and vitamins interact in various ways. For example, it is difficult for the body to use Calcium unless sufficient quantities of Vitamin D (particularly D3) are present in the diet. On the other hand, if the body doesn't have enough Magnesium, then the Calcium will "substitute" and there might be a deficiency of Calcium available in the body. Some minerals help regulate specific processes in the body -- Chromium is often considered to be important in insulin production within the pancreatic glands.

Vitamins and minerals can be obtained via meats and other animal products. However, since vitamins can be weakened, or destroyed, by cooking, meats are not a preferred source of vitamins.

Fruits and vegetables really shine when it comes to providing the body with vitamins and minerals. They are usually cooked less, or left raw, and this means the nutrients are left intact for the body to use. In addition, the organic compounds which contain the minerals are thought to be better utilized by the body than those from an inorganic source.

Alas, fresh fruits and vegetables are quite expensive on a per-calorie basis. Canned produce are less expensive but often have reduced nutrients due to the canning process and, in the U.S., are often compromised with added salt and sugar. Frozen fruits and vegetables are often the best balance of nutrition and cost -- but, of course, requires a freezer for storage.

One of the biggest challenges in food preparation of fruits and vegetables is countering the media advertising for processed foods. The more processing, the more profits. The more home processing (or preparing a recipe, if you prefer), the more time needed. We will go more into these trade-offs in the next blog, which will be a summary of nutrition with an emphasis on economics.

31.   The Science and Economics of Nutrition, Summary – 12/13/2012

We've seen that nutrition involves bringing into the body a full set of "building blocks" such that the body can build, repair, and maintain itself.

Starvation occurs when the body does not have enough calories to maintain its "operational needs" -- not enough fuel to keep going. When the body does not have enough calories, it first burns stored fats and then starts burning proteins -- which include muscles and organs (such as the heart) and eventually causes death.

Malnutrition occurs when the body does not have enough of all of the different "building blocks" to create, repair, and maintain the various components of the body. This is particularly devastating to the young when they are initially forming the body -- it can cause long-term effects. (In adults, temporary malnutrition can be recovered from unless it lasts too long.). Illness and inability to perform daily tasks well are often the outcome of malnutrition.

In the U.S., we are fortunate that private charitable food banks and government programs make starvation almost non-existent. However, malnutrition exists to a considerable degree with a greater concentration among the poor.

There are four components of achieving good nutrition. These are knowledge of good nutritional needs, action taken based on that knowledge, time, and money. Most of the focus is on knowledge -- but many educational programs in school attempt to avoid science and rely on "rote" formulas. This lack of foundational understanding of nutrition makes the task of achieving good nutritional balance difficult in an atmosphere of mass media marketing. False, or misleading, claims are easily accepted. Sometimes it causes rote formulas to be followed such as "red meat is bad" without understanding WHY red meat CAN be "bad".

Money directly enters into nutritional decisions. Good nutrition is more expensive than poor nutritive, high calorie choices. Given a sufficient budget, however, it is possible to provide good nutritional meals but it requires time to plan, choose, and prepare good meals.

I have never encountered a "30 minute meal" that I can prepare in 30 minutes. A parent who is working two (or three) part-time minimum wage jobs does not want to allocate the time -- there is homework to work with, houses to clean, medical appointments and soccer games to juggle. Well balanced restaurant meals are expensive but fast food alternatives are widely, and energetically, marketed and sold to the public.

Frozen vegetables are more nutritious but take more time to prepare than canned. Given a choice between a $1 apple and a $1 candy bar -- which do you think most children would choose?

32.   What's the Fuss about GMOs?
– 2/18/2013

There has been a lot of flurry about Genetically Modified Organisms (GMOs) in the press and Internet. What is a GMO and what are the concerns about it? GMOs are actually the tail end of a sequence of food modifications -- the process is the least "natural" and the most uncertain as to long-term consequences. The series begins with natural hybridization, leads through human hybridization, and continues to GMOs.

- Natural hybridization. Plants and animals change over time. Through a process of "natural selection" and "spontaneous mutation", life changes to adapt to best survive in a particular environment. Sometimes this combination of processes is called "evolution". A "spontaneous mutation" isn't anything menacing or bad -- it just means that the "child" is significantly different from the "parent". If the change brings advantages then the "child" is more likely to survive. If the change brings disadvantages then the "child is less likely to survive. This is "natural selection" and it applies to all plants and animals.

- Human-directed hybridization. Plants and animals are naturally diverse. They have slightly different characteristics from each other. By choosing life that has attributes that are "desirable" to propagate to the next generation -- mixing and choosing -- new combinations of attributes will emerge within the offspring.

The difference between this and natural

hybridization is that the new attributes are not
normally chosen based on the plant surviving
without human intervention. In fact, the opposite is
often true. Human hybridized life often requires
ongoing human intervention. This may include
more water than is naturally available within the
region. It may require special weeding and
chemical support. If left alone, without human
support, it will often "revert" back to the varieties
that best survive.

- A GMO takes this choice one (considerable) step
  further. The actual seeds, or eggs, are manipulated
  to add or remove genetic material from one form
  of life and integrated together. The goal is to make
  the genetic change inheritable from one generation
  to the next. One can look at it as "non-
  spontaneous" mutation. Desired attributes may
  include "better" flavor, easier transportation,
  longer lasting after harvest ("shelf life"), faster or
  greater growth, or greater production (more fruit
  or milk produced, for example).

There is nothing specifically bad about this -- it is
just hurrying nature along. However, many of these
changes are highly unlikely to ever occur
spontaneously. Some genetic material from
animals may be added to plants or vice versa.

The primary warning, or fear, from GMOs is that,
by introducing life that would probably never occur
naturally, there is little knowledge of what the long
term interactions within the ecosystem, or between
producers and consumers, will be. There may be
little difference between the genetics of a
nutritious plant and that of a slow-acting poison.
Studies of new organisms rarely are long-term,
checking effects through the generations.

Another problem is that GMOs may be patented.
The courts sometimes take a contrary view of this
by applying existing patent logic. Existing patent
logic is based on the idea that patented
ideas/materials can spread only by being "taken"

or specifically reproduced. Thus, plants that contain the patented genes have "stolen" the new material. While this follows existing patent logic it does not apply properly to living material that can naturally spread. It should be treated as "invasive" or an "infection" where the GMO is actually "attacking" the non-modified life. Patent law needs to be updated to existing realities.

GMOs are not inherently bad. They **are** inherently a change and changes take a while to determine benefits or risks. GMOs already are in wide use within human food. Some countries require GMOs to be labeled as such, feeling like the public should be aware they are part of a long-term study of effects. However, due to the widespread use of GMOs, food producers often fight against this notification. Specifically, corn is often a GMO plant and, in the U.S., corn syrup is used in many food products. Most food (including sweetened carbonated beverages) would need GMO labeling because the product includes GMOs.

33.   Smoke Gets in Your Lungs – 2/22/2013

In the present day, there's a lot of controversy about smoking. Old substances (tobacco) are in decline and are often under public disdain. "New" ones (marijuana) are in ascendance and are becoming more acceptable and may even cross the line back into legality (it's not really new and has a long history of various stages of legality -- see the Wikipedia entry Legal History of Cannabis).

However, because of its controversial legality and use, the health aspects of marijuana smoking have not been pursued as it has been for tobacco. There are also other substances that are sometimes smoked -- heated and inhaled or brought into the mouth. So, let's take a step back and just look into the smoke.

Smoke occurs when substances are burned. Many firefighter (and people trapped in fire areas) injuries are associated with smoke inhalation. There are good (but highly technical) sources such as Wikipedia for articles on the various possibilities depending on factors such as heat, presence of other substances, humidity, and so forth. In this blog, we will concentrate on the tobacco and marijuana smoke voluntarily inhaled in relatively limited amounts by "smokers" and "tokers".

1. Smoke has four major aspects: carbon monoxide, tars and particulates, active ingredients, and additives.

2. Carbon Monoxide. This gas is created by incomplete burning of the material. It is primarily a danger to those "inhaling" as it is more lung-related. It is present in all smoke in varying degrees and is considered to be a poison as the body will absorb it and it

81

decreases general function and may cause death. It is absorbed into the blood cells and decreases the ability to absorb oxygen and, thus, reduces stamina and general ability to perform at optimum levels. There is little difference in the effects between tobacco and marijuana smoke.

3. Tars and Particulates. These are the visible parts of smoke -- if you can see it, they are present. They are what cause the darkened areas of X-rays of lungs. "Tars" are resinous substances -- usually quite sticky and they cause the staining of teeth (and lungs) and can act as a kind of glue in the lungs, reducing the ability of the lungs to absorb oxygen. The particulates can vary in size depending on the temperature of the burning material and the size of the parts being burned and can act as an irritant causing more mucous production which is a major cause of "smoker's cough". It takes about five years for the lungs to fully repair from the damage caused by these substances and is considered to be a major trigger of lung cancer. This area could use more study but it is likely that there is little difference between tobacco and marijuana smoke.

4. Active Ingredients. These are the substances that interact with other parts of the body to cause the effects anticipated by smokers and tokers. In tobacco, this substance is nicotine which has many effects but primarily acts as a stimulant. It also has a mild diuretic (body water removing) effect which may cause weight loss for beginning smokers (and cause some temporary weight gain after quitting). It is water-soluble and, thus, is ideally administered via smoking. Anecdotally, nicotine is considered to help general focus and mental activity but there are no controlled studies verifying this. Nicotine also tends to paralyze the cilia in the trachea (windpipe) and, thus,

may increase mucous retention and coughing. There are a few minor medical purposes for nicotine but it is highly physically addictive.

The primary active ingredient in marijuana is tetrahydrocannabinol (usually referred to, for obvious reasons, as THC). Its effects vary from individual to individual and is considered primarily a "psychoactive" (acting primarily on the central nervous system) drug. Common effects include increased appetite, decreased nausea and pain, and a reduced sperm production in men. It is not considered to be highly physically addictive although a varying amount of emotional or psychological addiction is possible for both marijuana and tobacco.

5. Additives. Since tobacco has been legal for many years and is a highly competitive industry, each company does what it can to both distinguish and "enhance" its brand of tobacco. Glycerin is normally added to increase shelf life and prevent the tobacco from drying out. Formaldehyde (think of frogs in jars in biology class) is added to make absorption and crossover to the brain more rapid. This increases the risk of physical addiction. Many other additives are also present -- from spices such as cinnamon and cloves to liquids such as menthol and other oils. Formaldehyde is known as a carcinogen and the burning by-products of the other additives have not been extensively tested. However, this is an area that is highly likely to contribute to act as cancerous triggers (especially for non-lung-related cancers). Tobacco is much worse in this area than marijuana.

Inhaling or not inhaling? As a water-soluble drug, the effects of nicotine are dependent upon the surface area times the duration of exposure. This is also true for other additive burn by-products. Thus, inhaling would be worse if holding it in the mouth was done for the same amount of time. However, this is not actually the case as cigar and pipe smokers tend to allow the smoke to remain in their mouths longer. Thus, there may be close to the same exposure for nicotine and additives. However, there would still be greatly reduced effects from the carbon monoxide and particulates. With THC, it varies upon the general environment. It is not considered to be water-soluble but it IS lipid (fats) and alcohol-soluble so, if taken with food or drink, it is probably absorbed as readily (or more readily) than bringing into the lungs.

34.   They DON'T Make Them Like They Used to – 3/9/2013

People sometimes say "they don't make them like they used to" -- usually referring to something that has broken down and has to be recycled, discarded, or replaced. This is a truth that has entered into a special category of use.

The reality is that the way things are made is always changing. Sometimes they are made faster, with new techniques. Sometimes they are made with new features and new technologies. Sometimes they are made, deliberately, to not last as long.

There are three general categories of change in how/why things are made. These can be called "marketing", "manufacturing", and "labor".

- Marketing. This is the active art of consumerism. This is what makes people want to buy something. The global economy is presently structured around consumerism. The rationale falls into what I would call the "three Fs" -- fashion, features, and failure.

  Fashion is a desire for something "new" for the sake of having something new. Change in styles are presented as being "better" than what currently exists. "Orange is the new Purple" (Purple having been the previous preferred color). "Chrome is in". Skirt lengths go up -- or down. Teak is the preferred wood. So, out with the old and in with the new (although, if you keep it long enough, the cycle will come back someday).

Features. New programs require faster computers or more memory. Faster speed requires new connectors and those old connectors won't work anymore. This hat has a higher SPF (Sun Protection Factor). The new game has more versatile character sets and better graphics. There is "improvement" but marketing works to move it from the "want" to "need" category.

Failure. I don't think that many manufactures REALLY design their products to fail (they rely on fashion and features more to entice you to get something new). However, they do have a desired lifetime for the product when they create it. Too short and you won't buy their products again. Too long and you won't need to buy their products again if fashion or features don't draw you away. And, within that designed product lifetime, it causes choices to be made in manufacturing. Given a choice between a less expensive part that lasts "lifetime + a little more" and a more expensive part that lasts "three lifetimes" -- they will make it with the less expensive part. So, manufacturing (next section) is designed with the projected lifetime in mind.

- Manufacturing. As mentioned in a previous blog, technology and manufacturing rely on a pyramid of tools and less complicated parts. Manufacturing a DVR requires laser technology, semiconductor technology, power technology, and so forth. What this means to the consumer is that the end product is made up of more complicated, but fewer, separate ("discrete") parts.

  It may be faster, smaller, and less energy consuming ("green") but it will also be more complex. Most technology is now manufactured largely by "robotic" technology with humans doing the specialized work that doesn't justify building a more-specific "robot". This is even true for things like clothing where the fibers are made, or spun, or extruded by factory processes and then machined/loomed/created by machines that may

86

only need humans to replace spools or to adjust things that have gone slightly askew.

In summary for manufacturing -- fewer, more complex parts that avoid human interaction in creation.

- Labor. This area directly goes "hand in hand" with manufacturing. In countries where labor costs are "high", it may cost more to repair something than to replace it. In countries where labor costs are "low", you can find items repaired in extremely ingenious ways (they weren't designed to be repaired) because the cost of repairing is less than replacement.

So, they DON'T make them like they used to and we probably don't want them to do so. What we do want is to have the items last as long as WE want them to rather than as the manufacturer has determined. Often, choices are available but we don't know what those choices include -- expected durability is not something that is advertised on the packing of products. We rely on consumer groups and other people's ratings to make choices. And, I guess, that is the best we'll do for a while.

35.   Metabolic Balance: To Lose or Gain – 3/19/2013

People talk about **metabolism** when they talk about weight gain or loss. Metabolism is actually the entire process of conversion of food into energy and building blocks. What they are truly talking about is "metabolic rate". Someone who is considered to be a "fat burner" actually has a high metabolic rate. Most people know other people who seem to be able to eat just about anything (and everything) and stay thin.

The formula for weight loss (or gain) is "calories taken in - (calories * metabolic rate) = excess/deficit of calories to maintain weight). The body is very smart and it does NOT like to lose weight -- it considers it a "starvation" situation. Thus, when a deficit of calories is taken in, the body will reduce the metabolic rate in order to maintain weight. While it is possible to lose weight by "just" reducing calories, your body will fight it by lowering the metabolic rate and increasing the amount of hunger you feel. Since the body fights loss of weight, it will also respond to lack of calories by burning proteins in addition to fats which will also tend to reduce the metabolic rate.

As mentioned, the body is very "smart" -- it knows how much food is needed to maintain weight and it knows what building blocks (minerals, vitamins, proteins, However, in current society, we have deadened our abilities to listen to our bodies -- ignoring what it tells us. It is deadened by the pace of our lifestyles as well as the increased availability of calorie-dense foods which appeal to our hunger triggers.

There are things that CAN be done to return control over our ability to listen to our bodies AND take control over whether we are at the weight we desire.

- **Exercise**. This is probably the number one thing that we can do -- both aerobic (body moving, heart rate increased) and non-aerobic (weights, muscles working hard) exercise are of use. Aerobic exercise on a regular, prolonged, basis help to increase the metabolic rate directly. Anaerobic exercise increases muscle mass which requires more calories to maintain and, thus, effectively increases the metabolic rate. Contrary to popular belief, exercise does not increase appetite although it does help us to listen to our bodies more carefully -- so, if we are hungry, we will feel hungry. Try to drink non-sweetened liquids first, followed by bulkier foods such as celery or carrots.

- **Meditative exercise**. While some aerobic or non-aerobic exercise may take place at the same time, the primary benefit is allowing us to get in touch with the needs of our bodies. Activities such as yoga or meditation, in general, fall into this class.

- **Slow down eating**. Take smaller bites and eat slower. This allows the time needed for the body to say "I'm no longer hungry". In order to avoid the lag between eating and satiation (lack of hunger) it is good to stop when one is just a little bit hungry.

- **Drink non-sweetened liquid**. Water is good. Teas and Coffees and flavored waters also work. Artificially sweetened liquids (and foods) do not work well as they tend to actually increase hunger. Filling the stomach with non-sweetened liquids helps to reach satiation quicker.

- **Spicy food**. The capsaicin found in spicy peppers has been shown to increase metabolic rate a little. While this is not highly significant, spicy peppers tend to be mixed in with other vegetables and other non-calorie-dense foods.

- **Limit calorie-dense foods**. We talked about the various components of a healthy diet over a series of blogs. However, calorie-dense foods (high fat, high sugar) are easy to eat "too much" of very quickly. Eat them in small quantities and savor them (eat slowly while enjoying the flavor).

- **Increase bulky foods**. Vegetables, high-fiber complex carbohydrates (oatmeal), and such will satiate with lower calorie foods.

The bottom line is to allow your body to tell you what it needs. This is very difficult in modern society with its fast pace and a heavily marketed variety of calorie-dense foods but, at the end, you have changed your lifestyle to enjoy the food you eat better and to regain control over your body.

36.   Analog and Digital Data – 4/21/2013

I thought that I would talk about digital media -- CDs, DVDs, Blu-Ray, and so forth. But then I realized that I really needed to first talk about what digital media are -- and that, in turn, means that it is important to talk about analog.

Analog data are a reflection of events that occur on a continual basis. Such things include time, temperature, sound, moving images, water flowing, and so forth. An analog watch is known by its "face" -- where the "hands" are located to allow a person to interpret the data (information).

It would be completely possible for a watch to have a single hand. All the information is present in the hour hand. However, it is difficult to "read" (interpret) the value with a single hand and, therefore, analog watches and clocks normally have a minute hand and may even have a second hand.

In my old university, they had an analog computer. Set up correctly, it would be able to be used to calculate an exact value for Pi. But this brings up further the problems with analog data -- being able to actually make use of the data in a precise manner. An analog thermometer can give a precise value but can a person really read it that clearly? Is it saying 98.6 or saying 98.53?

Digital data can only create approximations of continual information. There are a lot of non-continual data in the world -- particularly in the area of finances. However, when it comes to continual data, you are involved with sampling rates and precision. The sampling rate is how often you "mark down" the information. You take a sample of sound at 1 hour, 20 minutes, 15 seconds, and 180 milliseconds. You then take a sample of the sound every 20 milliseconds -- but, whatever interval you choose, you are also choosing to ignore the data that exists when you are not taking a sample. You will never really know what happened within that 20 millisecond gap. You can **guess** what it might be -- that is called interpolation -- but you cannot know.

The second part of digital data is precision. For money (or other non-continual data), the precision is self-defined by what exists (although other units may exist for formulas -- like taxes). For continual data, the precision is a choice. Do you record 98.5, 98.54, 98.536, 98.5359, or what? Once again, you lose data/information and your choice CAN make a difference if the data are used in a repetitive fashion (such as calculating trajectories for a space ship).

So, analog data are accurate but very difficult to interpret precisely. Digital data are an approximation but have ease of interpretation as a built-in aspect of the choices that are made.

And this leads into the next blog "What makes Blu-Ray (TM) Blue?

37.   What Makes Blu-Ray™ Blue? – 5/8/2013

In the last blog, I talked about the difference between continuous (analog) data and discrete (digital) data. One of the most popular, "hands-on", types of data that people use each day are that for audio and video.

In the case of audio, analog data are often considered to be the most "faithful" to the sound. Digital adherents say that the sound stays crisp and clear. They are actually both correct. When analog recordings are made they are able to reproduce all of the "between" sounds that are dropped during analog recordings. While one can debate as to whether it can be heard by most people, it does exist and, therefore, there may be a substantial difference even if only noticed by the subconscious.

Analog recordings primarily fall into two categories -- an engraved reproduction of the sound waves or a magnetic version. Each has the capability of continuous data recording. However, the use of such recordings requires destructive mechanical mechanisms to be "read" after being recorded. For the "engraved" (vinyl, records, wax cylinders -- yes all have been used) version this means a sharp object following the path of the engraving which will eventually start cause eroding the engraving and a deterioration of the sound. For magnetic versions, the media (tape usually) wears while being pulled and the magnetic material on the tape also gets worn by friction with the reading "head".

Thus, as time goes on and the recording gets used, the recording will get worse -- while, in general, a digital recording will stay the same. So the audiophiles and the digital adherents are both "right".

CHALLENGE: It should be possible to create a commercially viable analog recording medium that can be read non-destructively. With all the bright people and companies employing bright people this should be possible. Make it so!

As discussed in the previous blog, digital media (for audio and video especially) requires decisions as to how much data will be omitted. This is precision and sampling rate. For human speech, it is considered acceptable to take a sample 8000 times per second and the data can be recorded with the use of 8 binary units ("bits"). This means that a digital recording of human speed will require 480,000 bytes (8 bits), or 480 KB per minute of recording. In the case of "high fidelity" digital recordings of music, the sampling rate can be increased and the precision may also be increased. This ends up with a greater amount of data.

Currently, a popular way to record this data is on optical disks. The digital data are marked on the optical media with very, very small pits. A pit can be considered to be a "1" and a land (non-pit) can be considered to be a "0". Note that this is actually very similar to analog engravings except for the nature of the data. Please also note that the exact encoding is actually more complicated than I am saying -- check other sources for more precise descriptions.

A larger difference from analog data, however, is how the data are read once recorded. An optical disk makes use of a laser which can tell whether there is a pit or a land by the timing of the reflection from the medium. This reading is non-destructive and, as long as the optical medium is not otherwise damaged, should retain data unchanged for a long time.

We now enter into the third area of data recordings which is storage space. An audio CD makes use of a near-infrared laser (wavelength of 780 nm). This wavelength determines how dense the data can be placed on the optical medium. For an audio CD, using 780 nm wavelength lasers, about 737 MB (megabytes) of data can be stored in a single layer (a disk CAN have multiple layers with the laser reading separately from the different layers of the disk). Since this amount of data is considered to be around 80 minutes of music, we can see that, for an audio CD, each minute takes about 10 MB of data so the precision and sampling rate are much higher than considered acceptable for human speech -- greater "fidelity".

The wavelength of the laser determines density -- how "packed" the data can be. This limits the total amount of data in a predetermined physical size. One method of increasing the density is by decreasing the wavelength of the laser.

A Blu-Ray disc uses a "blue" (blue is officially considered to be 475 nm) laser with a wavelength of 405 nm. A single layer Blu-Ray disk can store about 25 gigabyte (GB) of data. For audio, this would be about 100 hours using the same encoding as CDs. DVDs use a wavelength of about 650 nm ("true" red).

38.   Simplified is Not Necessarily Better – The Tyranny of the Average – 5/23/2013

There is a strong tendency in our society to try to make things "simple". I don't know whether it is because we are so time rushed or because we have such an imbalanced system of education. Sometimes having it simplified works to a specific person's benefit -- sometimes it is quite unfavorable to a specific person.

Simplification is often linked closely with statistics. As Mark Twain is quoted as saying -- "there are liars, damned liars, and statisticians". Statistics are applied math so they should be accurate -- however, the actual use of the formulas and numbers are decided by people and people make decisions based on what they want to have be true (either consciously or subconsciously).

One example of this are the figures quoted by politicians during election time. "The average income tax went down during my administration". Hmmm. Well, this statement could be "true" if the average income also went down during the time. It could be "true" if the tax rates went down (this is the interpretation the politician probably hopes you will make). It could be "true" if tax rates went down for one segment of the people (usually the high-income group) but went up slightly for other segments (and this has been happening in the U.S. for the past 12 years or so).

Three different realities -- all "supported" by the same "facts".

One area that hits hard for me is the Body Mass Index (BMI) number. The BMI is a fast, easy, "simple" method to indicate whether a person is overweight. It is reasonably accurate (+/- 5% or so) for about eight out of ten people. Who are the other 20% of the population? They are people who are particularly tall (more than 15% above average) or short (20% or more less than average) or people who have large amounts of muscle tissue -- yes, the "fit" are most at risk from inaccuracies on the BMI.

All of this would be academic if it wasn't so easy to fit these simplistic numbers into other formulas -- such as actuarial tables used by insurance companies. So, if you happen to be a body builder then be prepared to pay more from private insurance companies (on life and health) for being "too fat". If you are very short, then you can be rather overweight and still have the insurance count in your favor. If you are very tall, however, then you need to be prepared to pay more once again. Oh -- and I forgot to add -- computerized dating systems tend to like to use BMI in their calculations so expect body builders to be matched up with others of ample dimensions.

The reason the BMI is used is that it is easy and cheap -- take your weight (in kg) and divide it by your height (in meters) squared and you have a handy, dandy, all-purpose number. In order to really find out an accurate number, it is necessary to find real body fat percentage. There are formulas that measure different areas on one's body and then use those in conjunction with weight to get a number that works for 95%+ of the population. But it takes more time and more time means fewer patients seen and that means smaller profits. It is also possible to do a submersion test (where your body is submerged into a tank of water to accurately determine volume) that is the most accurate way to measure density (weight divided by volume equals density).

These are two methods where the average can hurt those who don't fit. There are many others. But do your best to understand the implications of statistical statements -- it may not mean what it seems.

39.   The REAL Food Pyramids: Sustaining the Foundation – 6/6/2013

Several organizations around the world have attempted to create graphical representations of what we should eat -- sometimes called "food pyramids" (current USDA version is a "food plate"). At first, it looks like this should work pretty well since food percentages are roughly 65/25/15 (carbohydrates/fats/proteins). Unfortunately, when this translates to actual food, most food is composed of a combination of nutritional elements. All fats are not considered to be of the same benefit and the effect of carbohydrates varies immensely depending on included fiber and other nutritional building blocks. Thus, it is difficult to use a pyramid to represent food needs. People still will often think of these food pyramids.

However, there are true food pyramids based on the needs and abundance of life on the planet. These are sometimes called trophic pyramids or energy pyramids. At the foundation level of these pyramids exists life that uses the energy from the sun (directly or indirectly) to manufacture food and body. On the land, these organisms are broadly called plants. In the sea, they are broadly called plankton -- although phytoplankton are the specific ones which are able to perform photosynthesis (creation from light).

These foundation foods (or primary producers) are eaten by the "higher layers" of the ecological food pyramids. It is possible for any organism to make use of them directly. For example, whales may feed on krill which are considered to be plankton (although they, in turn, make use of phytoplankton). In general, the lowest level are directly consumed by the next most abundant form of life. In a food pyramid, the next "level" can be determined by either number or function. Another way of putting it would be to think of a cartoon depiction of a very small fish being eaten by a small fish eaten by a medium fish eaten by a giant fish.

The organisms of the next level are called primary consumers. Primary consumers eat primary producers. So, herbivores are a general class of primary consumers. Although we usually think of mammals as herbivores, insects may be herbivores and worms might be considered to be herbivores.

The following level, sometimes called secondary consumers, may be either omnivores or carnivores. That is, they may eat a combination of producers (plants/phytoplankton) and primary consumers (or other secondary consumers) -- or they may be strictly carnivores that eat only other consumers. The "highest" level (in terms of the pyramid) eats only consumers.

These concepts are discussed in various ways -- food chains, food webs, ecological chains, and so forth. In whatever way they are approached, there are producers, primary consumers, secondary consumers, and tertiary consumers. The primary producers are always at the base.

The base determines the overall capacity of the entire pyramid. Thus, it is enormously important to protect that base. In the sea and on land, the largest threat is pollution although global climate change will certainly affect it in various ways. Note, however, that pollution can be either an aspect of waste ("garbage", runoff from managed land, etc.) or deliberate (even if accidental) contamination by oil and chemical spills and use of various chemicals within the food and non-food production chains. Plastics are also a growing threat.

As omnivores, humans have the capacity to shift their herbivore/carnivore balance -- they can be primarily meat eaters or primarily plant eaters. A shift towards the lower levels can allow more food to be available for all.

40.   The Minimum Maximum: What is a Bottleneck? – 6/14/2013

When you have a system that is made of various parts, there will always be some part which limits the overall performance of the system. Within flow situations (water, gas, data, etc.) this is called a "**bottleneck**". However, this concept can be extended to many things that we encounter in life, so I am going to discuss a more general concept that I call the "**Minimum Maximum**".

For example, you see a really high-performance car on the road. However, the performance of this car is much worse than you know of its capability -- it takes several seconds to start after the light says to go, it cuts corners or moves across the dividing lines of the road, its speed varies on a constant basis. The reality is that the performance of the car is based on the best ability of the car and the driver. If the car is great and the driver's ability is "poor to middlin" then the car will only be capable of being driven poorly. The purchase of a car is based on capability to buy -- not capability to drive. So, the high-performance car is "wasted" with the not-so-good driver. The car can be called "overkill" because its features and performance cannot be used appropriately.

If a race-car driver is behind the wheel of a poor-performance car then she or he will only be able to drive that car to the best of its ability. It is optimum to "tune" the system. Good cars for good drivers and poor cars for poor drivers.

In the world of the Internet, this is more often called a bottleneck. Let's say that you have a broadband connection that can provide 30 Mega bits per second (Mbps). However, you have an older computer and it can only process data at a rate up to 10 Mbps. The ability to use data from the Internet will be limited to the 10 Mbps of the computer. Going the other direction, if you have a powerful computer but your access to the Internet is limited to 128 kilobits per second (128 kpbs) then you might as well get a slower (and less expensive) computer.

This would also apply to general gaming (not multiplayer and not connected to the Internet). If your disk drive can access data at 5 Mbps and your processor can only display at a rate of 3 Mbps then your disk drive is faster than necessary.

The problem with many situations for optimizing (or tuning) is that it is done in a multipurpose way. In the first Internet case, replacing the computer (or upgrading it) would increase the overall performance. The same holds true for the last example (upgrading the computer will raise overall performance.

I'm sure that you can think of many other instances. When my family goes off to Church, we often have to wait for the "slowest". -- that person is the Minimum Maximum.

Whenever two or more systems interact, the various parts will work at various speeds/capabilities. Similar to doing Least Common Denominator (LCD) problems in school, the "MinMax" is the way to determine what is slowing down YOUR system (and an indicator as to what might need to be upgraded first.

41.   BIG Data and Data Mining – 7/11/2013

In my household, big data is most directly related to the piles of LEGOs (or LEGO-system building components) that my boys have scattered around the house. Needles in haystacks are more often used as examples. My library of books around the house would be another example. In each case, big data basically means a lot of data.

A lot of anything, of course, is subjective. There are thousands of pieces of straw in a haystack. There are a few thousand books around my house. My boys have a couple of thousands of LEGOs. However, in the world of business (and surveillance) big data usually refers to hundreds of thousands (or even millions) of records -- each of which may have many minutes (audio) or many members (items sold in purchase records or words in emails, for example). Big data is just a way to describe lots of data.

Data mining is the process of finding that special yellow 2 by 2 LEGO in the pile, or finding the needle in the haystack, or finding a specific audio record that talks about things that are considered suspicious or dangerous.

Data mining has three basic components -- collection, storage, and analysis. These are not necessarily discrete stages but we'll discuss them separately (calling out exceptions).

As evidenced by the physical examples at the beginning of this blog, big data has always existed. Consider the stacks of paper birth certificates, or other historical documents that exist and which may need, from time to time, to be searched. The ability to effectively handle, and use, big data has gotten much easier since electronic formats have become standard.

- Collection. Collection usually occurs at the time of transmission (when the originated data is moved to a destination). This might be a phone call. It could be at a point-of-sale (POS) cash register after the order has been finalized. It might be the registration record for a class. Collection may either occur at the intended destination (the company invoice/purchase order database) or via interception. Interception is where collection occurs somewhere other than the intended destination -- "wire tapping", people looking over your shoulder when you enter your credit card security information, and so forth.

  Collection can occur anonymously or personalized. Personalization basically means that the record is associated with a corporate or living entity. In the case of a sale at a grocery store, the data will be associated with that store (and, possibly, that cashier and cash register). If you use a credit/debit card or a store "club" card, then the data can (and probably will) be associated with the person in addition. Generally, anonymous collection is considered innocuous while personalized collection is not. This does not mean there are not "legitimate" (proper, honorable) reasons to collect personal data but it does mean that the person may have concerns as to the purpose and safety of the data.

- Storage. This always occurs at some point. However, it may be transitory if the data are removed upon receipt and analysis. Consider a "normal" phone call. The audio message exists (and is stored) from the origination (talking) until the receiving person analyzes it. If the message is redirected (to voice mail, for example), intercepted, or copied, this may turn into a permanent record requiring long-term storage.

Transactions (purchases, registration, email correspondence) where the data needs to be used in the future are almost all "permanently" stored. Of course, they can still be deleted in the future -- but, without advance knowledge of when, or if, this will occur they must be considered permanent.

- Analysis. This can occur during the process of collection or it may occur later (after storage). Anonymous data is often analyzed statistically. How many of product X were sold by store Y in city Z? How many of product X were sold in state B? How long is the average voice call within a state? Trends can be analyzed over time. Store Y in city X sold NN of product X at price B. They sold GG of product X at price C (can be used to determine overall profit using margin versus quantity sold). Product F sells very well during the time period D through G but not very well in period H through M (seasonal item to be stocked differently depending on time of year).

  Analysis can also be personalized. Customer ABC buys a lot of product F. Product G is similar but there is a greater profit margin on G -- send Customer ABC coupons for product G to get them to start buying product G on a regular basis. Or Customer DEF only buys product F if the price is below $ZZ.ZZ. Customer BEF is now buying baby products -- notify baby supply companies of contact information.

  Finally, analysis can be triggered. Surveillance can use trigger words, or sequences of words (either written or audio) to divert records to further analysis. If you start buying diabetic-related foods and medicines, the data CAN be forwarded to your insurance company (and yes -- if the data is associated with you, then they CAN find your insurance company).

Big data does not change the stages but it does change the methods. There will often be multiple layers of analysis so that each step reduces the number of records to be analyzed. Analysis upon collection will specifically affect the manner in which the data are sorted and stored. And so forth.

People usually don't object to anonymous statistical analysis. They may start feeling threatened with personalized statistical analysis although they may also benefit from the results.

They often will feel threatened with triggered analysis because their "private" data are being used without explicit permission and can be used to exploit the data in some way. In addition, triggers can lead to false conclusions quite easily (you were actually buying diabetes supplies for your great Aunt, you have been reading a book about bad thing XXX and were discussing it with a friend). Big data methods are particularly susceptible to false initial triggers (although, hopefully, further analysis will filter more appropriately).

42.   Body Sensors – What Your Body Tells You – 11/24/2013

Our bodies have the capacity to tell us a lot about our health and what we need to do to feel our best. Alas, we also have the ability to ignore these signs and an aspect of our modern society is to encourage us to do so.

When people think about "sensors" they usually think about electronic automation. There is a sensor to tell your thermostat whether to turn on the heater or air conditioner. There are sensors in ovens and toasters to indicate when proper temperature at appropriate times has been applied. There are sensors in our cars to indicate proper fuel intake and when to shift gears and even to apply brakes or throttle. We have similar things within our bodies. Much of robotics is concerned with getting machines to be able to do the same things we do every day.

The first sensors that come to mind for people is our " five senses". These are usually listed as sight, hearing, touch, taste, and smell. Some people add a "sixth sense" to indicate information we take in that are not easily linked to the five physical senses.

However, we also have a lot of internal sensors -- primarily associated with the way that our brains are able to interpret specific signals. We can tell, via pressure at specific points, whether we need to urinate or defecate. We can tell if our stomachs are adequately full. The sense of cold and heat can easily be fooled because it is associated with the way specific nerves under our skin react to temperature differences. Sometimes these interact with certain organs -- such as our inner ear -- to tell us whether we are level or spinning and help with the ability to move smoothly.

The third category of sensors is difficult to understand fully. Our brains have access to much information that requires a host of tests to determine externally. They have access to insulin levels, to endocrine levels, to the amount of oxygen being carried by our blood, and to the levels of neurotransmitters and other chemicals in our brains and bodies. Much of the time, our bodies work with this data automatically by use of the "brain stem". However, it is possible for people to access this information consciously and actively apply responses.

So, what does all of this information tell us? It tells us when we are hungry or full, whether we are hot or cold, and whether the food, drink, or other substance we might bring into our bodies is good for us or not. It tells us whether we need to use the restroom. It also tells us whether we are tired, sad, happy, stressed, excited, and just (all in all) how we feel.

Consider now the various items that often exist in our homes, or in the supermarket/pharmacy, or being advertised as services for us. Many of these exist because we do not pay attention to the information our bodies tell us. Why not? The pressures of a time-obsessed society cause us to eat quickly (not giving time for us to listen to body signals), and schedules tell us when we can eat/drink and do other bodily functions. The allure of a "quick fix" stops us from adopting a lifestyle where we get the proper exercise and sleep. Calorie-dense food is easily available and our bodies did not develop to allow for such. Plus, we often feel that it is a "reward" to do things that our bodies do not want or need -- that extra large dessert or an "extra large drink".

It isn't easy to change and our economic society does not encourage us to change. However, if we allow ourselves to listen to what our bodies tell us then we can be healthier and happier.

43.   How Sweet it is: Sugars and the Body – 12/15/2013

Our bodies, when we pay attention to them, include a group of tastes. These are usually referred to as sweet, sour, bitter, salty, and umami. They work by the chemicals activating specific sets of nerves on the tongue. Bitter tastes are associated with poisons and sweet with high-energy foods. It is rare, in nature, for one taste to dominate and this causes a huge set of possibilities. Also, the "flavor" of foods is how the brain interprets the taste, smell, and texture of the food in combination.

Since sweet tastes are an indication of higher-energy (calorie) foods, our bodies tend to favor sweet foods. Sweet tastes also activate a swallowing reflex within the mouth. (If a person is dehydrated but has trouble swallowing, try adding a single teaspoon of sugar to a glass of water.)

In the past, a "taste" for sugar has not been a problem as, within most of history, getting enough food has been a much greater problem than eating too much. With modern food processing, concentrated sweetness is a cheap method of making food more appealing and, thus, has become a problem for many people.

Natural sugars are a group of carbohydrates called saccharides. These may be monosaccharides which include glucose (a "blood sugar"), fructose (a "fruit sugar"), and galactose. Disaccharides include sucrose ("table sugar"), maltose ("grain sugar"), and lactose (a "dairy sugar"). I put these common referents in quotes because, although associated with these types of foods, the sugars are not exclusively in these foods. Other substances can also activate the sweet sensors of the tongue. These include a group of chemicals called glycosides, some proteins and amino acids, and even some inorganic compounds.

All fruits and vegetables include sugar as it is a by-product of the process of photosynthesis (conversion of water and minerals into food using solar energy). Some vegetation is considered a primary sugar source because the concentration of sugar is sufficient to warrant extraction and can be used directly as a sweetening agent for foods during cooking. Three of these are sugarcane, sugar beets, and Stevia leaves (which includes a glycoside rather than a saccharide)

Historically, natural sugars have been used as sweeteners -- adding to food to make them more enjoyable to eat. It is possible, by the process of extraction, to increase the percentage of sugar by eliminating the fiber (pulp) and leaving a solution of sugar and water (still including other water soluble vitamins and mineral compounds). A final method of concentration reduces, or eliminates, the amount of water and non-sweet compounds until only relatively "pure" sugar remains.

From the body's point of view, the unprocessed sugars are what the body was designed to appreciate. Eating an orange, combined with the pulp, is healthy (in moderation). Orange juice, with the pulp extracted, is much less healthy -- and a tablespoon of sucrose is the least healthy. Our bodies were not designed to deal with "pure", refined, sugars and making use of such within a diet can cause various problems, including an overwhelming of the pancreas causing diabetes.

In order to combat problems associated with natural sugars -- including high calorie intake and tooth decay (the sticky sugar can remain on the teeth as food for bacteria) -- humans have created "sugar substitutes". Many of these substitutes activate the sweet sensors on the tongue to a higher degree than natural sugars. This means that a much smaller amount may be used for equivalent sweetness (increasing profits and decreasing any caloric intake that may still exist).

This sounds like a win-win for producers of food as well as consumers but, as we will cover in the next blog, fooling "mother nature" can cause the body to react in ways that are not easily foretold.

44.   What's In a Name: Sugar and Marketing – 12/21/2013

When it comes down to looking at the way that words are used in marketing, the use of sugar is a prime candidate as an example. Have you heard of these terms? Unsweetened, no sugar added, sugar-free, naturally sweetened all seem to suggest a healthy drink or food but you had best examine the ingredient list carefully. What do these terms (and others) really mean? This is a situation where sometimes the literal meaning is usually the one that is "true" but most of us hear what we think it implies rather than what the word says.

Let's start with the word unsweetened. It is not used consistently even in the market. Sometimes it is used to mean "without sweetness" -- such as unsweetened tea or coffee. However, it is also used for situations where added ingredients include sugar but no separate refined sugar or sweetener is added. For example, tea with cream is a sweetened drink because cream includes various sugars (primarily lactose). Unsweetened cereal means no refined sugar was added to the mix but almost all grains include sugars (maltose, fructose, and sucrose).

I used to buy a slice of "no sugar added" apple pie at a local restaurant. I love to cook and bake (it's really an at-home chemical laboratory) and know that it is possible to make an apple pie without any added sugar (one does have to do something to "draw out" the moisture from the apples, however, or it will be quite dry) because the apples have enough sugar within. But, it turns out that, at this restaurant, they actually make use of Splenda (sucralose-based) which is an "artificial sugar substitute". So, their definition of "no sugar added" really means "no caloric natural sweeteners added". I guess that it doesn't have the same ring to it -- but it is a lot more direct.

"Sugar free" is one of my least favorite marketing phrases. Unlike "unsweetened" which, at least sometimes, means no sweeteners are added -- it almost always means artificially sweetened. I guess that "chemically sweetened" doesn't have the same marketing pizazz as "sugar free". I keep hunting for a real sugar-free drink but water seems to be the only one that can be really trusted.

Finally, there is the term "naturally sweetened". Generally, this does have a meaning -- no refined sugars are added and no artificial sweeteners are included. This does not often mean without considerable sweetness as it usually means that extracted fruit juices are used in combination with the "primary" flavors. For solid foods, it means the same but the added sweetness comes from the other ingredients (like the tea with cream) in the mix.

So, how do we determine what is actually in the drink or food? Like most foods, one has to learn to read the labels (in countries where nutritional labels are required). First, read the list of ingredients. Ingredients that end with "alose" (NOT "ose" which is at the end of most natural sugars) or "itol" are likely to be artificial sweeteners. Natural sugars end with "ose" or are described with everyday words (sugar (sucrose), corn syrup (maple syrup is the only other included syrup of which I am aware is used). Check the order and frequency. The highest percentage come first but if the list includes sugar AND corn syrup AND fructose then the total might very well be first or second highest amount -- another marketing trick to shift the order of ingredients and help you to think it has less sugar.

After checking the list of ingredients, one has to look at the nutritional label. In the "carbohydrates" section, it will be broken down into dietary fiber and sugar. The sugar should correspond to the natural sugars in the ingredient list. When you add up the amount of sugar and dietary fiber, it will usually be less than the total amount of carbohydrates. The difference between these two amounts indicates the amount of more "complex" carbohydrates.

Carbohydrates are a classification of food elements that combine carbon and Hydrogen-Oxygen (think "hydration" -- or water added -- although hydrates are not quite the same as adding water). As from the previous blog on sugars, dietary carbohydrates can also be grouped into saccharides.

114

The simple monosaccharides and bisaccharides are given the name of "sugar" while the polysaccharides are sometimes called "starches". Starches require the body to break them down into simpler molecules before using (as sugars). At any rate, the difference between the total of sugars and dietary fiber and total carbohydrates indicates the amount of "starch". As you can read from my earlier blogs on nutrition, the healthiness of starch depends on the mixture -- the ratio of dietary fiber versus sugar and simple starches should be kept high.

So, let the "buyer beware" -- the names used (and with many other aspects of life) have multiple meanings and uses. Marketing terms are used to make the consumer interested in the product -- not to inform. I really will talk about differences between "natural" and "artificial" sweeteners in my next blog but I thought that the use of marketing terms and sugar was important to understand first.

45.   "Real" or "Artificial" – 4/13/2014

First, I apologize for not getting back to my blog for a while. I have moved cross the U.S. and taken a new job that is quite a bit different from my previous job -- so it's been easy to get distracted. I'll try to post more regularly again.

There is a tendency for people to talk about food items in terms of "real" or "natural" versus "artificial". In reality, there isn't such a dramatic division between the two. Sucrose is "natural" in that it occurs, without chemical manipulation, in nature and in food items. But, table sugar is far from "natural" as it is necessary to do a lot of refining to have it available in the form that we use it. However, although it is not "natural" nor is it "artificial" as the chemical substance is not different from that found in life.

From the other direction, a chemist can duplicate a chemical compound found in nature. There is no difference between it (which is "artificial") and that which was "natural". Many flavorings, used in food, are of this nature. But there is actually a difference -- the one found in nature is mixed with many other flavors, textures, and other compounds (including inorganic fibers). Nature is rarely "pure". Sometimes this means the laboratory-created ingredient is "better" and sometimes it is not.

From the previous blogs, we have tried to decide what is the reality behind the names used in marketing of sweeteners. Once again, we find that it is a "gray" area -- some "natural" sweeteners are rarely used as found in nature. Other "artificial" sweeteners may be present in inert (does not interact with other substances) forms -- or they may be metabolized (broken down into building blocks by the body).

So, which is better? In answering that question, I will first say that I am not a food chemist, researcher, or any person who has a degree in a related area. I am a generalist.

My general feeling of analysis is based on experiential analysis -- what happens, what can be observed, how is it used? In other words, if it walks like a duck, quacks like a duck, and lays eggs like a duck then it might just very well be a duck or, at least, something you can call a duck.

When the body tastes something sweet, there are a number of reactions that the body undergoes. This is as a reaction to the quality "sweet". The body will react to this "sweetness" regardless of the source of the quality. Sucrose, Fructose, Sucralose, Stevia, and so forth will all make the body react because it is reacting to the defining quality of "sweetness". We can observe the body salivating. We can observe the swallowing reflex.

We can **not** easily observe the internal reactions -- and this is where it is difficult to compare and "prove" results. Although it makes sense that the body will react internally to "sweetness" the same independent of the source -- I cannot prove it. Let's say that it is true -- what does that mean? Mostly, it means the body's metabolism will start trying to process the substances with which it associates sweetness. Insulin will be produced and gastric juices will be increased to break it into smaller building blocks and to store energy.

So, the body does all these things based on "sweetness". What happens when it is NOT the substance for which the body has developed these reactions? What happens when insulin is released and there is nothing for it to react against? What happens when the metabolism tries to break down an inert substance? What happens when the body says "absorb" and the substance cannot be absorbed?

I don't know the answers. I can make guesses but I do not know the answers. But I am rather certain that I do not want my body to be used as a test bed to determine the long-term reactions and effects. I know that there are various undesirable effects from refined table sugar -- but I know them and (although not in the refined version which has only been easily available for a century or two) it has been in use by people for a long time.

"Natural" is not always good. "Artificial" is not always bad. But, in the area of food items, one can also look at it as "what is the body used to", 'how was it designed".

118

That's my two cents on the subject. I'll stick to regular sugar. What do you think?

46.   Race: The Invented Divider – 6/8/2014

The definition of race is primarily that of running as the word comes from the Middle English word **ras** meaning "to rush". However, although it may be one of the lesser definitions of the word, if one talks about "race" -- without article or pronoun -- the definition that comes to mind is often that of a division of humanity into different groups.

This idea first came into being to separate groups based on observable physical traits in the 1600s and continued in increasing use, and refined definition, through the early 1800s. Although some of the scientists had neutral goals for the use of the division, it was primarily used as a method of justifying colonization and subjugation of one group over another. It is largely discredited as a useful methodology within science at the present time.

In the elementary school that my children attended, there were about 850 children. If you lined up the children according to skin pigmentation, you would have a long continually varying set of shades and colors from near alabaster to ebony. If you lined them up on a summer day, you would get a different ordering of people from that on a winter day. The same thing holds true for color of eyes, or hair, or width of nose. Each physical characteristic varies on a continuous stream -- although there are certainly areas of the world that are more homogenous (similar between individuals) than others. This is why it has been abandoned by science -- it makes no sense to have discrete classifications.

Just because race has no reasonable definition does not mean there is not racism -- bias and prejudice based on observable (or known familial) physical traits. Racism, sexism, religionism, and other bias/prejudice are forms of **xenophobia** (fear of "the other"). It has existed since the beginning of recorded history and most likely since the rise of consciousness. The only cure for the syndrome is knowledge -- understanding of "the other" such that the similarities become more obvious than the differences. During periods of antagonism and preparation for war, differences are accentuated (made to seem greater) by governments in order to inflame xenophobia and the inclination to distrust and fear.

Some people say that an ism can only take place by the group in greater power toward the group having less power. Thus, there can be no sexism by women having bias or prejudice about men and there can be no bias or prejudice by pigmentally enhanced people towards those who are pigmentally challenged. This makes as much sense as the original xenophobic reaction. If there exists bias or prejudice based on an observable (or known association with a group who DOES have such observable) physical traits then the relation of power makes no difference.

In summary, the best way to work with, and reduce, xenophobia in all of its forms is knowledge and understanding.

47.   How Long Will I Live: Life Span and Life Expectancy – 6/28/2014

There is no consistency in estimates of how long people have lived throughout history. This is largely because, prior to around 1500 Common Era, birth and death records were rare -- usually only available for royalty or others who had influence and power. The everyday person's birth and death were remarked upon only by friends and family.

In general, however, it is considered that the overall life expectancy has increased over the years. One set of estimates indicates around 25 to 30 years old in BCE, rising to 30 to 40 years old in the 1500 to 1800 and then ballooning up past the 1800s to current world expectancy of 65 to 75.

Life expectancy is a statistical measurement as to the AVERAGE life span for a larger population. This number can vary between regions of the world, countries, or even counties. In the previous paragraph, I was talking about global numbers. An individual's life span is dependent on different factors. Some of these factors are not under anyone's control, some are "per chance", and some are voluntary risks.

The primary factor for life span is how long your ancestors lived -- your genetic heritage. Robert A. Heinlein did a great job going into this in his book *Methuselah's Children*. This is the baseline -- something that we presently cannot change and which gives the maximum time our bodies have to be around without becoming zombies. Some people believe that this can be extended by various means but, in my opinion, it is really a matter of eliminating the many factors that can shorten this period -- the maximum has not truly changed.

We have succeeded in helping to prevent some events that shorten life -- which is why our global life expectancy has increased. One of the biggest boosts in overall global life expectancy has been from medical advances that have decreased infant and mother mortality. If a quarter of all children die before they are two years old, it decreases average life span (and population life expectancy) considerably. This is also a large part of why the life expectancy of women is now higher than that of men and why it used to be the other way around. Other medical advances and general sanitation have been the other primary method to avoid life shortening events.

So what are the common life shortening effects? **War** (and murder) is a huge one and one which historically has taken a greater direct toll on the expectancy of men. **Drought** and **famine** change localized life expectancy. Lack of nutritious food early in life can also affect health later in life even if food is then available. Death by **disease** has been reduced by immunizations, treatments, sanitation, and recognition and isolation. Finally, death by **accident** is with us and seems to be impossible to totally avoid.

Voluntary risks do not really add to the life shortening lists -- they just make them more likely to occur. Smoking, for example, can increase the chance for disease if one is genetically likely to get the disease -- a trigger effect rather than a causal one. Enjoying a dangerous hobby -- parachuting, mountain climbing, car racing -- can increase the chance for accidents.

Note that deliberately avoiding risks can sometimes actually increase shortening effects. For example, the overuse of antibiotics is increasing the likelihood of disease by making the diseases stronger as well as decreasing our immune systems' ability to fight disease. Living "in a bubble" (isolated) may decrease the chance of accidents and disease while one is "in the bubble" but it makes us even more vulnerable when we are no longer isolated.

So, aside from choosing our parents (which isn't possible), we can best increase life expectancy by having cooperative societies (lack of war and murder), producing and distributing nutritious food and healthy water, building up strong immune systems, and making reasonable choices to avoid preventable accidents and diseases. We still won't live forever but do we want to?

The next blog will address the downside of living a long life.

48.   Why be Healthy? – 8/16/2014

We've all read articles, or watched programs, where someone who was in a healthy "lifestyle" keels over and dies in the midst of jogging. The publisher of Prevention, a health-oriented magazine, died during a taping of a television show. A man smokes three packs of cigarettes a day and lives to 95 while someone who has exercised, eaten well, and never smoked dies of lung cancer at age 40.

Why be healthy? This is a serious question. There is very little correlation between specific lifestyle changes and length of life. (Mental serenity and positive attitudes do seem to promote a longer lifespan.)

Each study that emerges seems to indicate something different -- butter is bad, butter is good. Don't eat fats, eat only this kind of fat. Eat more carbohydrates, don't eat any carbs. Weight training is a solution to being healthy, aerobic exercise is the only thing that is important. If you try to follow along with the latest directions indicated, your body will seem to be at the end of a yo-yo. I loved the line in the movie "Sleeper" where they told Woody Allen to smoke a cigarette because it was one of the healthiest things for his body. I doubt that's true but the movie does point out that "knowledge" isn't static.

So, why be healthy? I would categorize these reasons into three categories -- triggers, quality, options.

- Triggers. As discussed in the previous blog, many of the health-oriented studies are **NOT** describing "causative" situations. Having a high-fat diet will not clog your arteries. Salt does not make your blood pressure rise. Smoking does not cause lung cancer (if so, then **every** person who ever smoked (tobacco or other substances) would get lung cancer).

However, if your genes say "I have a tendency towards high blood pressure and I cannot process excess salt" then a high salt diet may **TRIGGER** health problems. Since, for people with these genetic tendencies, it is possible to reduce the chance to get diseases which can decrease lifespan -- these are active measures one can take to live longer.

If you do NOT have these genetic tendencies then change of behavior may not make any difference at all. As mentioned above, however, this year's orange may be last year's black. Your best reference as to what is likely to make a difference to your body is your family health history.

- Quality. There isn't a chart or a set of numbers that says whether or not you are healthy. However, you can listen to your body. If you break out in a sweat every time you raise your hands above your head to put a dish away into a cupboard -- that is not a good sign. If you are out of breath after you have walked from the store to your car holding a bag of groceries -- that is not a good sign. If you have indigestion after most meals and are taking half a bottle of antacid to calm it down -- that's not a good sign.

  You very likely know what you should do to make it better -- it's just not easy with bombardments of advertisements for excessive, low nutrition food or a multitude of ways to be entertained with only a finger or two moving. Reading a book is passive but it means that you are controlling your own stimulus rather than being controlled from the outside. Make conscious choices.

  Just ask yourself -- was the sixth slice of pizza

worth the pain of your stomach afterwards?

- Options. What do you like to do? What do you WANT to do? Are you able to do it (or learn to do it)? Working towards a healthy lifestyle expands your options. If you are badly overweight, there are many things you cannot do. If you cannot breathe, it affects your stamina and limits the length of activities. If you are physically weak, additional limits exist.

It is easy to be a slug on the couch. You might even live a long life doing such. Is it what you want?

What makes it harder for you to do the things you feel are healthy for yourself?

49.   The Economics of Supersizing – 8/31/2014

I have to be careful when I talk about economics as it is such an all-pervasive subject that it is easy for me to lose focus. I consider it to be "applied sociology" -- or a measured way of evaluating how people interact and value each other within society.

Once upon a time, during a telephone interview with Google, I talked with them about how I thought Google was in a fantastic position to create an interlinked database of products, employment, and salaries. As only one example, such a database, and associated tools, could be of enormous help in figuring out how to migrate from a fossil fuel economy to a renewable fuel economy while minimizing the effects on the economy and individual workers. (Later, with 35 years of software architecture, programming, and managerial experience, Google called me in to interview for a marketing position -- they definitely have a sense of humor.)

See how easy it is for me to lose focus!

In the area of focus for this blog, supersizing involves a combination of total profits and perceived value. Perceived value is a subjective matter -- it depends on the individual and their history. In the US, it is considered to be of greater value to get more food for less money per amount -- in spite of the fact that the greater amount is unneeded and ends up being waisted (misspelling intentional). In most European countries, quantity does not enter into the equation for value as much as quality. In some other countries, it is a sufficient struggle to just get enough to eat.

When a product is sold, it is sold at a specific price. This price can be determined in one of two general ways. These are basically "cost plus" or "demand pricing". With "cost plus", the price is determined by a specific amount added to the cost of producing the item (including all overhead such as building costs, utility costs, storage, labor, and inventory loss). So, if a thingamabob costs $1 to make, store, sell, and so forth and the company wants to make 20% profit on selling thingamabobs, the price will be set at $1.20. With "demand pricing", the price is set to the highest amount that will lead to the greatest total profit. This is a bit more complicated.

"Net profit" is the difference between all the costs associated with making and selling something and the amount for which it is sold. In the "cost plus" example, there is a net profit of $0.20 or 16 2/3% (20 divided by 120). In "demand pricing", net profit is determined in a similar fashion except that the goal is to maximize the total profit.

In order to maximize total profit, the goal is to sell the MOST possible at a specific net profit such that the total amount is the greatest. For example, selling 1000 of something that has a net profit of $0.20 will give a total profit of $200. Selling 500 of something that has a net profit of $0.50 will give a total profit of $250. So, even though you are selling less, you end up with a greater amount of total profit. But, if you get especially greedy and start selling something a a net profit of $1 and only sell 100, you will end up with only $100 profit.

The practice of pricing for "demand pricing" is an art and involves marketing (convincing you it is something you want), branding (letting you recognize the product and make positive associations that increases its perceived value), and competition.

If you have a product that is desired by people and you are the only one who makes the product then you can demand the greatest amount. If you have a product that is made by many different companies and there is little perceived difference of value, then you enter what is called "commodity pricing" which usually has small net profits per item and requires mass production and sales to be profitable.

So, we come down to the area of supersizing (finally, you say). Supersizing (in the US) does two things -- it increases the perceived value and it increases the net profit (it MAY also increase total sales because of the increase in perceived value). Let's say that you sell a tidbit that has $0.50 costs associated with what goes into it (raw, or pre-processed, food ingredients), $0.30 labor, $0.50 overhead (such as building, heating, lighting, franchise fees, etc.), and $0.30 for sales (marketing, "free" toys, posters, advertising, etc.). You then sell the tidbit for $2, giving a net profit of $0.40/item (or 20%).

If you can convert that sale into buying something bigger -- let's say twice as big. then the only thing that you have increased is the costs of what goes into it. [There is, admittedly, a little more overhead concerning storage of more stuff but that is often balanced with a reduction in cost of buying raw materials.] So, rather than $0.50 of stuff going into it, there is $1 associated with the costs. You then sell the item for $3 and you make a net profit of $0.90/item (or 30%). If you make it three times as large and sell it for $4, you would make a net profit of $1.40/item (or 35%). This is how supersizing translates into FAT profits (OK, I admit it, I like puns).

In summary, as long as people see greater value in buying more food for less per amount, it will be difficult to persuade companies to not supersize as this is an easy way for them to achieve greater profits. The only route is to change mindset to demand greater quality rather than greater quantity.

50.   The 1% and the Former "Trickle Down" Pyramid – 9/16/2014

Once upon a time, when there was a lot less disparity between the rich and the poor in the US, an idea was proposed to justify giving more tax breaks to the rich. This idea was called the "trickle down" theory. Note that this blog concentrates on income inequality -- wealth inequality has parallel workings.

At heart, the trickle down theory was an economic pyramid and is the basis of the concept of regulated capitalism. However, the main word in that sentence is **regulated**. The idea is that the people with the most money generate more money which is distributed (in lesser amounts per person) to a larger set of people, who then generate more money which is distributed (in lesser amounts than the people of the second level) to an even larger set of people. At the bottom of the pyramid are the people who are unemployed or are working for whatever they can get paid and not die.

The original pyramid for the "trickle down" idea worked rather like this (note that these numbers are all just examples):

1 person earns $1,000,000/year
3 people each earn $500,000/year
6 people each earn $300,000/year
10 people each earn $200,000/year
25 people each earn $100,000/year
55 people each earn $40,000/year

This forms a pool of 100 people. In total, they earn $11,000,000/year. The top earner gets 25 times as much as the lowest paid earner. and the top 20 people (20%) make 57% of the total money (leaving 43% to the lower 80 people (80%). The top earner gets about 9% of the total.

However, in order for this distribution to hold, it is necessary to have laws and regulations that keep redistributing money to the rest of the people according to their wealth. In the 1980s, it became politically popular in the US to think that if the rich were allowed to accumulate more money then there would be more money to distribute -- or "trickle down" -- to the rest of the population. That started a process of steadily increasing tax loopholes and favored treatments, lower (if paid) tax rates, substantially lower wages (based on pre-inflation 1985 dollars), and concentration of wealth which led to a new structure such as the following (once again, these are made up numbers -- the real ones are different but not better):

1 person earns $5,000,000/year
3 people each earn $300,000/year
6 people each earn $200,000/year
10 people each earn $125,000/year
25 people each earn $60,000/year
55 people each earn $22,000/year

Once again, this is a pool of 100 people and, together, they earn $11,000,000/year.

However, this time the top accumulator (no longer calling them an earner) gets 227 times as much as the lowest paid earner. The top 20 people (20%) have increased their total to 76% of the $11,000,000 but look carefully (this type of statistical use is often in political advertisements) -- the 19% below the top 1% are actually earning LESS than they used to. The top accumulator now controls 45% of the total money pool.

This is a situation where the top accumulators redistribute the earnings of the lower rich, middle class, and working poor to give to themselves. I call this distribution the "*splash over*" economic theory -- or a vivid, real, example of unregulated capitalism. You fill up the top and some of the excess splashes over to the bottom.

This was the situation in the "Gilded Age" in the 1800s. It was shifted, for individuals, with reforms such as the creation of income tax toward the turn of the century -- and it was shifted, for businesses, with "New Deal" reforms that came out of recovery from the Great Depression.

And, at the root of it all, the voters carry the responsibility.

51.   Why Aren't Studies Steady? The Problems with Studies on Humans - 9/27/2014

If you don't like the results of a study ... wait for the next study. It appears that the results from studies involving people vary drastically from study to study ... and they do. At one time butter is bad and margarine is good and then, later, margarine is bad and butter is better (not necessarily good). Fats are bad. No, the right fats are good. Olive oils are the right fats. No, more polyunsaturated fats are even better. Why do studies that involve humans vary in their results so much?

There are quite a few reasons why it is difficult to have consistent results from studies of humans. Some are inherent problems. Some are political problems. And a large number of problems arise from the way studies are reported in the media rather the actual study. In other words, a study may be done very well and present results that are interesting but not conclusive -- but some specific parts are taken up by the media as "startling results". What are the problems with studies?

- People are not mice. Many studies on health effects are done with mice, or guinea pigs, or monkeys, or chimpanzees, or some other more easily studied animal. In addition, many studies may make use of one gender but generalize to both genders. It is rather obvious that, for the best results, use of humans of the appropriate category must be used. Why isn't this the case?

Cost. People want to be paid (in money or value) to participate in studies. Animals can be purchased -- and, until some group starts recognizing what is being done to them, can be treated with the least care needed.

Morality. Except in certain situations (such as Nazi Germany where psychopaths had full permission to experiment) it is not acceptable to put people's lives at risk. This is associated with control groups where they are NOT treated the same as the group which is being tested as well as with the groups being treated with undetermined results.

- Patience. It takes TIME to determine what long-term results are. Of course, it doesn't take much time for an immediately lethal poison to be known but most substances aren't as immediate in effect. Taking longer periods of time means results are delayed. It also increases the costs of the study.

Two types of studies of this nature are longitudinal and cross-generational. One examines individuals for substantial portions of their lives and the other examines the effects from parent to child to grandchild. Humans have pretty long lives (unless stopped by disease, accident, or violence) and this also leads to use of shorter lived animals as subjects.

- Control groups. A control group is simple in concept but much harder to create and use. The idea is that one group has the variation (tries a medicine, eats a food, does an exercise, endures environmental conditions, ...) and the other does not. Comparing the results of the two groups is hoped to be able to isolate the effects of the specific effect being studied.

A huge difficulty is that it is impossible to prevent the confusion of *combinations* of variables. You are testing item **A**. It turns out that **A** does one thing in the presence of items **B** and **C**. It does another, different, thing in the presence of items **C** and **D** --

and it may even do something else in the presence of **B** and **D**. If **B**, **C**, and **D** are all known and defined then useful results may still be obtained -- but often they are not. These combinatorial variables are unknown -- but they can drastically affect the results. Two of these variables involve environment and genetics.

Environmental variables. What is the effect of electricity in a house or city? In modern society, it is impossible to eliminate -- and, if taken to a part of the world where electricity is NOT present, other variables will exist. What is the effect of plastics? What are the effects of pesticides, hormones, or antibiotics in the food? While these can be minimized, they cannot be eliminated. What about specific pollutants in the air? And so forth.

Genetics. Leading up to the next bullet item on correlation versus causation, I once read a statistical study on smoking and cancer rates in various countries of the world. It turned out that countries with the greatest amount of smoking had among the lowest amounts of cancer. The effect of smoking depended upon the population. (It didn't get a lot of media attention since the outcome was not politically popular.) A homogenous (identical in nature) population is needed for studies and sets of identical octuplets are hard to find.

- Correlation versus causation. This is understood by scientists doing studies but easily distorted by politicians, business owners, groups, or media people who have a bias towards a particular result. Causation means the variable **causes** the result. If I hit my toe with a hammer it will be bruised (or broken). This is true if *any* toe hit by *any* hammer causes these results. The effects can be changed -- a hammer hitting a toe that is shielded by a steel-toed boot is NOT hurt (but it also means the toe of the foot is not actually hit).

Some people are allergic to monosodium glutamate (MSG). Some are not. So, MSG AND

141

allergic people cause a particular effect but MSG AND non-allergic affect does NOT cause the effect. This is verifiable contributory causation.

Everyone who drinks water will eventually die. Does drinking water cause people to die? No (unless, of course, the water is contaminated), but this is the type of (often statistical) result that biased groups enjoy mis-interpreting and reporting.

- Binary Results. People like simple results. They like "yes" or "no". They do not like long combinations of possibilities. So, a report that says "butter is bad" is much easier to distribute than a report that says "butter, in combination with lack of exercise and excessive refined carbohydrates and a genetic tendency towards high cholesterol, can contribute to high blood pressure".

Properly done studies rarely have simple results.

So, in summary, it is difficult to create a useful, consistent study on the effects of anything with people. Even if done properly, it is difficult to give results without also giving all of the controlled variables along with the result.

52.   Fracking: What's the Big Deal? – 11/1/2014

Fracking is talked about a lot nowadays -- mostly heatedly and almost always in a binary fashion (YES/Good or NO/Bad). What is fracking all about?

When we think of drilling for some liquid substance, what first comes to mind is something similar to putting a straw into an underground reservoir. Some may include the idea of a bunch of very wet sand, where the liquid is dispersed through the area but freely available.

Both of these were probably true for the earliest wells. A hole was pierced in the soil and it either hit an area of liquid (water, oil, also natural gas -- which is not liquid but, in this case, has the important properties of a liquid) or it was able to find an area that the liquid could collect in from the surrounding area. For water, this is called the aquifer. For water and other "liquids" (including natural gas), this ability to collect from the surrounding area is making use of a quality called "permeability" -- the ease of movement of liquids/gases through solids.

Pools are pools. It is a matter of finding them and using them. Permeability decreases as the area gets deeper because of the weight of the earth above. Greater pressure lowers the permeability of the surrounding area.

As we make use of more of our natural reservoirs of water, oil, and other liquid/gas resources, it is harder to find easy-to-use locations. Isolated pools become deeper (or underwater) and smaller (making them less economically possible to exploit). Natural permeability occurs more towards the surface and is harder to find. Thus, there is the desire to go where we have not gone before (national parks, wilderness, and beneath beds of water) and to promote artificial permeability.

Finally, you say, we get to fracking. Fracking is the process of breaking up the surrounding rock and creating permeability where none (or little) existed before. Early fracking mainly used explosives to break up the ground underneath. Although it usually worked, it was not very effective because of the irregularity of the fissures (or cracks) created and because, if it was deep, those cracks were likely to close back up from the pressure.

The modern practice of fracking usually involves hydrologic pressure. A liquid (often water) is injected into the well under great pressure and this causes the surrounding area to crack. Additives are put into the liquid to help keep the cracks open (thus, allowing permeability) after the pressure is removed. One method is called "acid etching" where an acid makes "grooves" in the cracks as they expand leaving little furrows when pressure is removed. This is done to make access to reserves more possible.

There are two primary problems with fracking. First, it requires a lot of liquid to achieve fracking. While that didn't use to be looked at as a problem -- with global climate change and uncertain icepacks and rains and overused aquifers it is quite a problem nowadays. I don't know the exact proportion of water to newly accessible liquid (you wouldn't normally use this technique for water wells, obviously) but there is a lot of water used. In addition, the additives put into water may not be either environmentally or health beneficial and those additives will find their way into the surrounding area.

The second problem is that the cracking of the earth is indiscriminate -- it does not JUST allow the desired liquid/gas to pool together. If there are pockets of water and gas or oil nearby, the fracking is likely to allow the mixing of such -- and causing health and environmental problems with use of the water.

The first problem might be addressable with new techniques (air fracking, perhaps, with environmentally safe additives) but the second problem is inherent to the process.

The only real method of eliminating the need for fracking to access dwindling natural reserves is to use alternate energy/base material sources.

What are your thoughts about fracking?

53.   What's In a Name: The Politics and Reality of Global Climate Change – 11/8/2014

When scientists discovered that the average temperature of the oceans was increasing, it was picked up by the media as "global warming". While I don't know whether this name was embraced by the scientific community or not, it wasn't a bad name -- at a scientific level. A much better name would have been "global ocean warming" -- for reasons to follow.

A name is a powerful thing -- attitudes, and historical associations, come along with names. If a highly anti-patriotic thing is NAMED something patriotic then it is easier to associate it with positive, patriotic, meanings. This is just how our brains work -- we associate names with other things connected with the names. Politicians, Public Relations people, and advertising agencies make use of this to a great degree.

If the word "global warming" is used then it can be MIS-used. If there is a huge snowstorm, it can be used as "evidence" that it is "obviously" NOT warming all over the globe. The more precise, and accurate, the word that is chosen, the more difficult it is to skew the interpretation. As mentioned earlier, it would be more difficult to misinterpret "global ocean warming" -- since a snowstorm is not immediately connected to the ocean (although it is actually true that the ocean warming might directly help the snowstorm to happen).

Of late, the dialog has mostly changed to that of "global climate change". This is associated with the **effect** of "global

147

ocean warming" -- but it seems to have been easier to migrate to this phrase than to expand upon the "global warming" media phrase. It is a reasonable phrase and much more difficult to subvert by politicians as there IS (unfortunately) increasing indications of global changes in the climate.

The other part of the "discussion" (actually more of a taking of sides) -- beyond whether or not global climate change is happening -- is whether it has been caused by human activities. It will be impossible to totally prove this as that would require comparing two parallel worlds -- one world where changes in energy use and other human activities took place in a timely manner and comparing that to our current world.

When historical evidence is examined, it shows how, and when, such global changes in climate have occurred before. This evidence does give us some ideas as to how various components (Carbon Dioxide in atmosphere, Ozone levels, water levels, average temperature, etc.) work together to change the climate and it does indicate how ongoing, present, changes are likely to affect climate. Politicians are correct that climate change has occurred many times over history. However, the **rate** of change of factors that have been occurring over the past 100 years has only been found in association with large scale catastrophes (huge volcano eruptions, widespread biologic changes, and so forth).

This rapid change is the scariest part of global climate change. Humans are very adaptable. In fact, a great name for the species would be homo adaptabilis. If a desert turned into a wetland over a period of a couple hundred years, people would adapt. In fact, much of the Sahara Desert was once a tropical forest at one time (but the change was much, much slower). If the coastline disappears under water an inch a year, people will adapt. That is our current goal -- to slow down the rate of global climate change to allow people to move from one place to another, to allow changes in basic crops grown in an area, to allow people to change housing and energy use, and so forth.

But, all of this is still an exercise in the dangers of allowing the media to choose a phrase to describe something. Insist on precision in descriptions. Refuse to use labels that are obviously inaccurate -- reverse it by saying "the so-called xxxxxxx". When a debate arises where one side is labeled as "pro-AAAAA" then insist that the other side be called "anti-AAAAA" rather than "pro-BBBBB"

What misuses of labels have been the most upsetting for you?

54.   Economics and the Meaning of Money – 12/6/2014

I usually look at the economy as a form of applied sociology. Money is only worth something if people believe that it is worth something. This applies equally to gold and jewels as much as it applies to pieces of paper with people's pictures printed on it. In a similar fashion, money is distributed according to the rules (explicit or implicit) that people decide upon.

A barter system works when each person (or family) is capable of doing most things needed for survival on their own. They then trade things that they have in excess for things that others have in excess. I give you an extra chicken and you give me a bushel of potatoes. I give you a length of material that I have woven and you give me a chair. Barter is a mixture of labor, materials, difficulty, and time combined into value.

When each person can NOT do most things they need for survival, the barter system becomes very inconvenient. It is necessary to keep records/charts of equivalences. One type A chair is equal to two meters of cloth. Two type B chairs are equal to one type A chair. Ten chickens are equal to one meter of cloth. This complexity arises out of the need for each family unit to trade for many different types of things. Once this happens, the next step is to equate the value to something in common. Ten seashells represent the value of one chicken. A meter of cloth is equivalent to 100 seashells. Every item of value can be represented by a certain number of seashells. This representation of value is called money.

Once the value of work and things have been "abstracted" into money, it is very easy to lose sight of real value. The work done by an experienced, talented teacher is probably worth more to society than that of a software developer -- but the software developer probably makes a higher salary. In "capitalistic" societies, the control of money is considered to have value in itself. That is, if I possess one million seashells then I no longer have to produce anything of value myself -- the circumstances (earned and saved, gifted, or inherited) of having the seashells allows me to distribute some portion to other people who then produce the actual value (plus more for me to hold).

Within "new age" philosophy, it is popular to think that the economy is no longer a "zero sum" game. That is, each and every person, can earn as much money as she/he wants -- that there is **not** a "fixed pot" of X seashells in the pot and each can have as much as they want without reducing the amount that others are able to have. That's a happy philosophy but is it true?

Although money makes lots of games possible with the distribution and use -- the basis of money still goes back to production and use. If 100 people each want a fish but there are only 50 fish then the value of each fish will rise until the 50 people who most want a fish have them and the other 50 do not have them. If 100 people want a fish and there are 1000 fish, then the value should (the concept of money makes direct value difficult if not impossible) be equated to that combination of labor, materials, difficulty and time mentioned above. An abundance of resources (fish) causes value to go back to basics.

Our global economy makes distribution of resources extremely unequal. Most people estimate, however, that there are enough resources (food, labor, energy) to support everyone currently on the planet. The fact that that does not happen is a problem with distribution and allocation. But, there is still a limit. Perhaps at twice the population there would NOT be enough for everyone (in an ideal world). This argues that it is a "fixed pot" -- there is a limit of resources to be distributed. In order to eliminate the fixed pot, it is necessary to get rid of the limitations of resources.

Is there a way to eliminate the limitations of resources? I will look at that possibility in the next blog.

55. Money as Energy: Increasing the Pool of Money – 12/29/2014

In the previous post, I talked about how money is basically an abstraction of the combination of resources, labor, and energy. We are fortunate that we do, presently, have more than adequate amounts of each. Distribution of such, however, is very uneven and, thus, causes areas of poverty, famine, and other physical and social lacks.

I ended the previous post with the idea that -- although our current problems are more concerned with distribution rather than actual shortages -- the New Age idea of an unlimited pool of money is not currently a reality. Is there anything to be done about that? Is there actually a way that everyone can have more (even with distribution problems)?

To address that question, it comes back to the three components of money -- resources, labor, and energy. It also requires a fourth "catalyst" which is technology. By using technology, energy can be converted into additional resources and increased labor availability. This argues that energy is the prime limiting factor within economics.

We can look around at the world and see how the availability of energy (applied via technology) has increased the "wealth" of the world. Farmers, via the use of equipment (using energy and technology to create and energy to keep active), can produce much greater amounts of food than what one person working the ground with manual labor can do. Harvesting of material resources -- trees, ores, fish -- are possible on a much larger scale than a single person could do making use only of manual labor (allowing a hand-built boat and fishing equipment).

The above paragraph indicates how energy (with technology assistance) can increase the amount of labor. It does NOT increase the amount of resources. But the amount of food for people has been increased -- isn't that an increase in resources? No, it isn't -- because the ecological pyramid has not changed. The amount of base-level food has not increased. The plankton, plants, and other solar-using food plants have not increased. The labor has been used to change the varieties of food harvested and the distribution of the food (from other animals to people). In fact, due to pollution and other side-effects of application of energy to increase labor, the total amount of food resources may go down (decrease in sea life in general, decrease in fish population, decrease in non-human animal population).

Can energy increase resources available to us? Yes, in two ways. The first is a continuation, and expansion, of what we presently do -- redistribution. We find other, more energy intensive, methods of accessing resources. However, this often has negative environmental effects and is also just speeding up the use of resources. So, although it increases resources available on a short-term basis, it does NOT increase the amount of resource. A second aspect of this (still redistribution) is to bring resources from other places -- the asteroid belt, for example, is a potential area from which to redistribute resources.

The second method of increasing resources requires much higher levels of energy. Besides the potential of alchemy (changing one element into another -- possible with huge amounts of energy), there are many <u>endothermic</u> reactions possible with increased energy available. Endothermic means "requiring the absorption of heat". Thus, it is possible to convert raw elements into more complex molecules and, finally, into "organic" materials needed for human eating, or use for furniture, or such. This is actually a metamorphosis of resources and not an increase -- but it's "close enough" for our uses.

So, with energy, the pool of "money" becomes bigger. Distribution remains a major problem. A larger problem is making sure that the energy is renewable -- we do not want to empty the bank as that would cause widespread catastrophe for the existing economy. The other problems aren't directly concerned with energy-as-money but are related to social, and environmental, responsibility for using it in a life-affirming way.

56.   The Media is What Lies Between You and the Information – 1/25/2015

The dictionary tells us that the root origins of the word *media* is the plural of medium and is from the Latin word for middle. The plural word media has been taken over to indicate a particular type of medium -- associated with mass media (television, radio, printed or texted material). Thus, within this context, there may be a plural of the plural -- medias.

No matter how you pull the word apart, it still indicates something that is between. Roots of the concept are really part of what is now called *social media*. This might be surprising to people who have grown up in the Internet era but not at all surprising to people who grew up in a small town. In medieval times, there was often an official person -- the *town crier* -- whose job it was to stand on a corner or before a building and call out information considered to be important by the person who paid the town crier. This was often a method of the government to tell people something. More important, both in medieval times as well as small towns, was the *town gossip* who was an individual through whom information passed from people and groups all over. I am certain that, even in cave dwelling times, there was always a particular person who found out information and passed it along.

The primary difference between mass media and social media is that mass media is more of a one-to-many spread of information while social media is a many-to-many. One source of information is taken by an individual (or group of individuals) and passed along to a large number of people and that is mass media. Or many people provide information which is collected by one (or more) person that is passed along to as many people as who will listen. In today's world that collection point can be human or electronic.

All of these are good methods of collecting and distributing information -- but they do not guarantee that it will be **GOOD**, or valid, information. We see this currently in discussions about existing events. We are often divided -- and very firmly divided -- because the information sources (and associated media) are very different and the people either do not have the time, energy, or desire to research the information themselves to determine what is real.

So, people decide who they trust and rely on that information. If the media are trustworthy, it works well. If not, they are the sources of the lies and rumors that damage, and sometimes cripple, people and societies.

The moral of the story is -- make sure you can trust the sources of your information and verify for yourself when you can.

What media types do you rely on for information and why do you trust them?

57.   The Interconnectivity of Things : Work – 2/7/2015

A popular topic nowadays is the Internet of Things (IoT). This talks about tangible devices that are interconnected via the Internet. I might discuss that someday. However, IoT is really a subset of the interconnectivity of Things (InOT). And that interconnectivity applies at many levels. Today, in continuation of my recent blogs on economics, I thought I would talk about how work/jobs interconnect.

One way to examine this is to put forth a possible change to the economy -- a shift from private transportation to public transportation (to save space later, let's call this "the Shift"). How would this affect the economy and how would it affect jobs and work?

The first thing that can be done is to list the things that are associated with private transportation.

1. Roads
2. Parking Lots (including driveways and garages)
3. Gasoline (assume have not migrated to electric cars)
4. Consumable parts (tires, windshield fluid, batteries, etc.)
5. Car distribution (including sales and transport)
6. Car manufacturing
7. Car repair
8. Car maintenance (including washing, upgrades, and so forth)
9. Car disposal

I am certain that this is not an exhaustive list and, as we will see, each item can be broken down into many sub-items. The Shift is also not a truly radical shift as most things associated with private transportation also exist associated with public transportation. Thus, it is a shift for reduction rather than elimination and creation.

Let's look at the first item on the list -- roads. First thing is that roads occupy space -- lots and lots of space. The Shift would not eliminate the need for roads but probably no roads would need to be more than 2 lanes (one each direction). In fact, many low usage roads could probably be a single lane with pull-over areas (similar to low-traffic roads in Europe). As a conservative estimate, let's say that, with the Shift, we could reclaim 80% of the land currently used for roads. (Skipping to item two, we can also probably reclaim 95% of the space needed for driveways and garages.) It also reduces the size needed for bridges and tunnels. What can this space be used for? Parks, farmland, recreation, pasture -- whatever space is presently used for.

One website indicates about 18,000 square miles (about 0.6% of the land area) are used for roads in the US. That site argues that that isn't very much but I would note that there are 9 states in the US that are each less than 18,000 square miles and it would be the same square miles as Rhode Island, Delaware, Connecticut, and New Jersey combined. The Shift would reclaim about 15,000 square miles or a little more than the size of Maryland.

Space was the first part of the first item. Roads also includes heavy construction equipment, labor (about 140,000 workers in the U.S.), concrete and asphalt manufacture and transportation, bridge and tunnel design, and so forth. Let's say that there are 200,000 jobs associated with roads and the Shift would eliminate the need for 150,000 of them. This means that 150,000 people in the US would need to find different jobs.

And this is just the very first item associated with the Shift. Every item above involves jobs, land use, energy use, and so forth. As a conservative guess, the Shift would cause the need for a million people to find new jobs.

So, yes -- many reasonable people would like to see the Shift, but resistance to the Shift is not just an automatic resistance to change. It is also based on the economic turmoil to families and economic infrastructure.

Once upon a time, I suggested a tool that would make analysis of such changes as listed above much easier. In my next blog, I will go into more detail of the design of such.

In the meantime, what effects do you see would come out of the Shift and would it be worth it?

58.   A Living Wage: It's Not that Difficult to Figure Out – 2/21/2015

It used to be that discussion was about the "minimum" wage. That was always a difficult discussion because it is totally subjective. If a person is starving and you agree to give them a sandwich and a glass of water if they work for you for eight hours then it is a minimum wage (anything less and they would die and be unable to do the work). Then there is the official "minimum" wage -- which is completely fictitious. People are paid less than the minimum wage all the time -- sometimes legally and sometimes not but it definitely is not the least amount of money people are paid.

However, when we come to the concept of the "living" wage, it is really easy. One thing to recognize up front is that there is not a single living wage. A living wage will depend upon the expectations within a society. It will depend upon the general cost of living in the area. It will depend on individual circumstances -- do you have children or others dependent upon you, do you have additional needs that others do not have (blind, deaf, mobility impaired, ...), and so forth.

So, there is not a single living wage for all people. But it is easy to determine. Add up the costs of everything it is needed to live over a year and divide that by the number of hours that are considered reasonable in your society. Let's put together a case example for an "average" city in the U.S.

There are a number of categories that MOST people would agree on. There are also a number of other categories that people would not agree on (is a phone required? is recreation required? is television required? is air conditioning really needed? is a personal car required? is it necessary to be fashionable? ...) Minimum requirements will include such as:

- Shelter
- Utilities (water, heat, sewage, power)
- Food
- Clothing
- Health-related Costs (incl. toothbrushes, toilet paper, clothes washing, etc.)

There's a certain range within each category that is required. Sometimes you might find a great deal on an apartment (or live with your parents). Sometimes you can find used clothing that is acceptable. Specific numbers can definitely be argued about and I won't say that you're incorrect. However, here are some (not the only) realistic numbers.

- Shelter -- a studio apartment; $900/month -- $10,800/year
- Utilities -- basics for a small apartment; $100/month -- $1200/year
- Food -- for one person, no fast food, no eating out; $8/day, $250/month -- $3000/year
- Clothing -- 3 pairs of pants, 2 underwear, 4 shirts, 5 pairs of socks, 1 pair of shoes, 1 coat -- $200/year
- Health-related Costs -- [# taken for an Affordable Cost policy for an unemployed single person] -- $340/month plus $40/month for medications/co-pays; approx. $4500/year.

This totals $19,760 for a year. For simplicity, round it up to $20,000. If we assume that working a 40-hour week for 50 weeks/year is reasonable then that is 2,000 hours. So, a "living wage" for a person with no special needs is $10/hour NET. I emphasize NET because this is what they have to have in order to pay for it all. If they have to pay country/local taxes or union dues or anything then that is added to the NET requirements.

166

So, we have determined a "living" wage. Even assuming that you agree with the above estimates the numbers can be moved around. If you qualify for food stamps, you might reduce your needs for food. If you can get subsidized housing, you might pay $500/month. But the foundation needs stay the same. Note also that there are no costs for childcare listed -- this is for a single person with no additional needs.

But society cannot afford to pay such!!! This is the statement that is echoed by businesses and wealthy politicians. It the next blog (hopefully -- I get distracted) I will discuss the realities of paying living wages.

Meanwhile, what things do you consider needed to live? Do you currently live on less? How do make it happen?

59.   Living Wages are Not Only Affordable – They Help Businesses – 3/16/2015

It is often said by spokespeople for businesses that "we cannot afford to pay our workers living wages". However, there seems to be no difficulty in paying for increased costs for materials, or energy, or advertising, or increased costs of real estate, or any other such item. As I discussed in my blog about "supersizing", there are a number of things that go into the cost of an item versus its price.

The composition, or gathering of different parts, of the cost of an item will vary depending on the item. Some things are "labor intensive" which means that labor costs are a higher percentage of the cost. Others are based on scarcity -- or an aspect of "we have what you want -- who is willing, and able, to pay the most for it". In general, for many items, the amount of labor cost within the total cost for things that are actively made by people is a minority of the cost -- call it 30%. For stores that have high "turnover" (things sold quickly and new, replacement, items put on the shelves for sale again), labor costs are much less (such as for mass merchandizing stores) -- perhaps 10%.

For our discussion, let's just say that labor costs are 25% of the cost of the item. Doubling the labor costs would NOT double the base cost of the item to sell. It just adds an extra 25% -- so the base cost is now 125% of the former price. Let's say that the retail price (price charged to a general customer) was twice that of the base cost -- or an extra 100%. This means that the price is 112.5% of the original price (100% original cost + 100% original profit + 25% extra labor costs gives 225% which is "normalized" (brought down to a comparison against 100%) to 112.5%.

Now it is possible (even likely) that the merchant might want to keep their percentage profit rather than the actual amount. So, in the above comparison, the merchant got the same amount of profit as base cost. If we increase the base cost by 25%, the total amount doubled ends up at 125% of the original price (100% of original cost + 25% extra labor costs is equal to 125%; doubled gives us 250% and normalized brings it back to 125%).

We can see that even doubling the labor costs does not add a huge percentage to either the base cost or a retail price without penalizing the retailer. It can be argued that a 25% increase is still something that people are not willing to pay. After all, people do comparison shopping and retailers have sales, and price cuts (temporary or permanent). If Item X is sold at one store for $1.25 and the very same item X is sold at another store is sold for $1 then many people will choose to buy for $1. What would make people able, or willing, to pay more for products?

The first reason is that the above analysis is a simplification. Labor costs are NOT the same as wages. Although the blog on "supersizing" uses labor costs as a lump sum, labor costs are actually a combination of wages, benefits, the cost to find someone to work at the job, training, and other matters. Thus, doubling wages does not double labor costs. In reality, it will reduce "turnover" within the workplace and reduce the amount needed to find people to do the job and the training. So, a doubling of wages may actually only cause an increase of 20% overall (these numbers are all examples but probably in a reasonable range) so the product would only cost $1.20.

The second reason is what do people do when they make more money? Well, hopefully they will save some more. But almost everyone would also spend more. The products may cost a bit more but the business is also creating more customers and a percentage will buy from their store.

A third reason is that it creates a positive image. I am sure you can think of a company who does not treat their employees well and relies on charities and the benefits paid by taxpayers to subsidize the wages of their employees. Similarly, we can also think of companies who pay their people more than what is "required" and are known for treating their employees fairly and well. Because of these three reasons (and other reasons) these "good neighbor" companies often make a better profit than the ones who sponge off of the taxpayers to increase the owners' wealth.

The last reason leads into a future blog (maybe the next one). And that is -- it isn't always a matter of "nice people finish last". The above three reasons come into play to help people who do the good, proper, thing benefit financially. Regulations also help -- because the companies who care about people (and environment, and health, and ...) are not penalized because they operate "on a level playing field". That is, if everyone is required to do something good then no company is at a financial disadvantage for doing what is good. Everyone has the same requirements.

Can you think of other benefits to a company for paying living wages?

60.   Regulations Create Level Playing Fields for Businesses – 5/9/2015

It seems to be fairly "normal" for businesses to complain about, and fight, every new regulation that is proposed or enacted. This isn't unreasonable as it will be true that a new regulation will require different procedures (and probably additional paperwork). However, that is not the same as saying that regulations are bad for businesses -- but it can be a difficult balancing act within the global economy.

Regulations are a way to tell businesses what practices are acceptable to the society in which they function. They fall into three general categories (actually, almost anything can be broken up into however many categories as are desired -- I am choosing three). These categories are economics, labor, and environment. There is also a fourth category which involves product regulations -- for the product quality and safety of the consumer but that does not directly apply to this blog.

Economic regulations involve the way the products of a company become part of the general economy. This will involve taxes. Generally, businesses want to pay fewer taxes and the general population wants them to pay more taxes. It will also involve tariffs -- both import and export. Tariffs are special taxes that are involved with the movement of products and money across country borders. This is a part of the balancing act.

Let us say that Country A, as part of the standards for their society, requires all businesses to ensure that any water used by the business to be cleaned to drinking standards before being released back to the environment. This requirement (or regulation) adds 5% to the cost of doing business in Country A. Country B does NOT have such a requirement and, thus, businesses can produce the same product for 5% less. This puts businesses in Country A at a price disadvantage. A tariff on products imported from country B gives the businesses a more equal competitive situation. (Note that the tariff does not help the environment in country B.)

Regulations may also be in the category of labor use. Minimum wage laws (or "living wage" laws whenever they start being enacted) say that people cannot officially be employed without a certain level of pay. Restrictions on number of hours worked per day, or week, directly affect the number of people employed. "Child Labor laws" restrict the age of workers and the number of hours per day that they can work at what ages. Mandatory sick days allowed (paid or unpaid) create a situation where workers are not compelled to work even when sick (this also benefits the general population when the food industry is involved). Vacation days, holidays, and other types of paid, or unpaid, absences help the overall health of the people who work for a business.

Environmental regulations are basically a matter of how businesses are allowed to affect the environment. Usually a person thinks of manufacturing companies for this. However, the requirement that a business have, and maintain, a parking garage would also be an example of an environmental regulation as it reduces the amount of land that cannot be used for vegetation. Another non-manufacturing law might be a requirement to turn off 70% of the lighting during non-working hours.

Of course, environmental regulations apply more directly to manufacturing businesses. It is similar to teaching a child to "clean up their own mess". A business would, naturally, prefer other people to take care of their messes. Note that not having environmental regulations does NOT decrease the cost to clean up -- it moves it from the business to the general public. In fact, it probably costs less for the mess to be cleaned up at the site of creation of the mess than after it has dispersed and damaged other parts of the environment.

It would be completely possible for a business to do everything well on their own initiative. They can treat their people well, be good to the environment, and be a good neighbor within their communities. There are many small businesses that strive hard to do such and other, larger, businesses that recognize that there are inherent benefits (lower turnover of staff, better public image, etc.) to do such. However, businesses that do NOT behave well can have financial advantages over their competitors -- and this does not help society as a whole.

Regulations provide a framework that is acceptable to the local society that allows businesses to compete without having uneven costs of providing services.

61. Our Bodies are Houses with Many Rooms – 5/25/2015

When we think of our bodies, we think of all of the various parts that built up from that original egg and sperm that started the process of dividing, growing, and specializing. From that beginning, we end up with a set of systems which allows us to stand, move, breathe, eat, and do the other functions of everyday living. These systems work together to make it possible for us to live and reproduce.

But, our bodies are NOT just composed of the cells that were created as part of growing and life. Our bodies provide an environment for many other living organisms -- some hostile but many of great benefit to our everyday lives. For many years, medical researchers have just noted the facts of their existence (and, sometimes, blindly tried to eliminate them) and had no real idea of just how they interact with the rest of our bodily functions. This is now changing and active research is being done. Note that this research is still young in experience and knowledge and many interactions and functions are still unknown.

An organism can exist anywhere within our body but we will focus on four areas that are of specific interest. These include the interior cell, the skin, the mouth, and the digestive tract. Within the interior cells exists organisms called the mitochondria (one of such is called a mitochondrion). While these "organelles" are an important mechanism to supply energy to the human cell (and, thus, to the human body), they also have many other functions within the body.

The mitochondrion reproduces independently from the other cells of the human body. It exists only within the egg (and not the sperm) and thus, inheritance of the mitochondria within a body is based on maternal lineage (mother to child). The separation between egg and sperm and its independent reproduction gives it a likelihood of being considered an *endosymbiont*. In other words, at some point in the past the mitochondria took up residence within human (and other animals) cells and act as cooperative parts of the cell.

On the surface of the skin (even the very "cleanest" of skin), exists a wonderland of tiny animals, bacteria, and other living creatures. The exact interplay between these creatures and their environment (the skin) is still to be fully understood but it is recognized that they do play a part in keeping the body from being invaded by unwelcome visitors. Of course, within this dermal ecosystem is the possibility of undesirable inhabitants -- such as fungus that creates "athlete's foot".

There is a large difference between the mitochondria living peacefully within the cell and the exposed environment of the skin -- a large, direct, interaction with the exterior world. Showers, baths, perfume, makeup, dirt, and so forth will affect the living environment on the skin. How does it affect it? We don't really know but two people (let's say they are identical twins to make this simpler) who have different exterior living environments, diets, and exterior habits will have different sets of creatures (or, at least, different proportions) on the skin -- which is likely to affect some aspect of health and protection from exterior invasion.

Our mouths harbor a large collection of bacteria and other living matter. This environment can be altered by both external and internal processes. Cleaning teeth and gums can reduce the amount of bacteria and their by-products. Blood sugar levels (especially relevant to those with diabetes) can alter the number and type of bacteria living in the mouth. (Note that brushing, diet, gum health, and blood sugar levels can all change the smell of your breath.) Many oral bacteria have potentially bad effects -- but controllable -- and, presently, few beneficial effects are known.

The area of symbiosis of greatest present interest is the world of the digestive tract. Although it has long been known that interior bacteria can be directly involved with digestion -- for example, termites have special bacteria that allow them to digest cellulose (wood fiber) -- it was not known that the same was true with humans and other animals. It turns out that the composition of the neighborhoods of our digestive system can affect hormones, immune system, metabolism, allergies, blood sugar levels (not just insulin and sugar), mood, and other matters.

Just how, and what, makes a difference is still unknown but the knowledge that it makes a difference is truly exciting -- in reality the addition of an entirely new bodily system. This infant study, of course, gives rise to much speculation and marketing of things believed to make a difference -- such as probiotics which are supposed to enhance positive groups of bacteria within the digestive system. Little is known for certain but it is an additional, important, area to understand for better health and activity.

62.   Water Distribution: A Climate Change Problem That Needs to be Addressed Soon – 6/6/2015

Climate change is just that -- a shift of weather patterns. It doesn't actually matter whether patterns change because of human activities or because of changes in solar flares or shifts in the magnetic core of the planet. It is thoroughly documented that the global ocean temperature is rising and photographs indicate an increasing loss of glaciers in polar regions as well as high mountain ranges.

Humans have survived over the years because they are great at adapting to different living conditions. Humans can adapt to the current climate change if necessary. The problem with the current shift is that (in geologic terms) it seems to be happening so quickly. Effects are being noticed within a single lifetime rather than after generations. One serious problem that is arising is that of water distribution.

Water distribution is very important for crop growth, animal husbandry, and basic survival. Politicians may hold snowballs for dramatic effect but, for every snowstorm, there is also a new heat wave and drought. It is likely that crops that have required a certain mixture of days of sunshine and amount of water will no longer have the conditions they require in that same region. So, shifts in climate brings along a need to change crop management and decisions as to what to plant.

Problems of water distribution can be classified into five categories:

- **Storage.** Much of our fresh water is stored in snowpacks and glaciers. During the spring, summer, and fall, they melt and provide new water supplies when rain is not present. As the glaciers of the Himalayas, Sierras, polar regions, and other

181

ranges disappear, they will not be able to continue to provide fresh water supplies. Alternate means of storage are necessary. Huge reservoirs will be needed to take the place of the snowpacks and glaciers.

- **Rising Sea Levels**. The amount of rising sea levels is unknown. If all of the snowpacks melt (including all of Antarctica) then a rise of over 200 feet is possible. This will drastically change our coastlines and many of our most populous cities are near the coastline. People will not just lose their beachfront properties but entire states and islands could vanish beneath the sea. These people will need to be relocated.

- **Redistribution**. There may be as much total precipitation as ever. However, the locations where it comes down (as rain or snow) will change -- weather patterns and local climates will change. As mentioned above, this will require shifts in crop plantings (note that some areas will receive more rain while others receive less) as well as general harvesting and crop planning. Denser areas of human population will also be affected with needs of water transport and storage.

- **Aquifers**. Much water is stored within layers of soil called aquifers. As droughts occur, or human population increases, water is removed from aquifers. This works fine as long as the amount removed is the same as that which filters back into the aquifers. In prolonged drought conditions, the levels of aquifers lower and lower. Wells run dry, deep rooted plants die, and the ability to bring the water levels back up start entering spans of decades. Aquifers are renewable but require management and continued rainfall.

- **Waste**. In many areas of the world, fresh water is in abundance. In other areas, it has always been scarce. These areas will change and waste of water in formerly abundant areas will need to be greatly reduced. Three to five minute showers will be needed. Ground cover that survive with existing moisture will be needed.

There is time to start planning for these needs and some may not be needed for decades but it is good to start the planning sooner rather than later.

63.   Processed Foods: The Good, the Bad, and the Ugly – 8/8/2015

The first step in processing foods is the act of harvesting. So, saying that a person doesn't like processed foods is a bit on the silly side. Picking an apple off of a tree or shucking an ear of corn is processing the food.

The next stage of processing involves changing the food physically. Food is cut up. It is ground. It is juiced. It is shredded. If it doesn't look the same as when it was harvested, then it has been changed physically.

Now we come to the part that involves chemical change. This isn't always what people consider to be chemical change. Drying something in the sun will involve chemical (and physical) changes. Fermenting a juice into an altered liquid that contains alcohol is a chemical change. Applying heat and cooking a food is a chemical change. Putting something in a different substance (even water) will often -- but not always -- initiate a chemical change.

Processing food makes it easier to eat -- or more interesting to eat -- or makes it easier to store for longer periods of time. These are not bad goals. So, what is the big deal about processed foods? Why would anyone object to processed foods.

As usual, the problems arise out of the details -- the hows, whats, and whys. Harvesting isn't usually considered a big problem -- unless it is done in a way that has "side-effects". For example, some foods are most easily harvested by starting a big fire and sifting through the remains. Other foods might be harvested by killing the main plant when only a small part is going to be used for food. Finally, there is the background issue (not to be discussed at length here) of production which is necessary for harvesting to occur.

Physical processing takes energy. If physically preparing food is part of the daily routine then it is a closed system -- energy given by eating the foods is used in the preparation of those same foods. However, when we buy something already processed then we are also requiring energy to be used for the preparation as well as transport. That energy is likely to include non-renewable energy sources.

Now comes the most controversial part of processed foods -- chemical changes. The previously mentioned changes would usually be considered "natural" (although they are not all likely to occur without some intervention by some other living entity). When "artificial" processing (as mentioned in another blog, these terms don't have as distinct of a difference as people often seem to think) is involved, the food will change in ways that cannot always be predicted.

Edible items (we'll continue to call them foods) can be produced from inorganic (not originally alive) components. During the processing of foods, additives can be included that change flavors (this can include spices and herbs -- but also inorganic chemicals) or make the food last longer (usually called preservatives), or become more addictive (added sugars, salt, certain chemical substances) to increase marketing and sales. These types of processing can make the food less healthy -- and this is the big deal about processed foods. Commercially prepared processed foods are often less healthy than fresh or personally processed foods.

So, why do people buy processed foods -- and why are people buying a larger and larger percentage of their food in the form of processed foods? I would categorize that in the regions of time, convenience, marketing susceptibility, and (for many items) monetary savings. Using processed foods saves time -- it reduces the time to prepare, the time to consider the recipe, and usually time to cook or put on the table.

In the case of processed fresh fruits and vegetables, you will probably be paying quite a bit extra to have someone process them for you -- but you may save, if you are single, because you are only buying the amount that you will eat. This is a cost trade-off.

In the case of stored (frozen, refrigerated, canned, or packaged) processed foods, it is often true that it will cost you less to buy processed food from the store. This is because the food processing company pays less for the ingredients because they buy in huge quantities and because the energy costs in labor and machinery are less than what it would take for you to prepare. The actual costs of the food may amount to 1/4 of what you would have to spend to prepare it personally. With the markups of transportation, storage, and profits at different stages, the final price is still often less than what you could realistically prepare. So, it ends up being a choice concerning time and quality.

Marketing susceptibility is a huge topic (and one on which I have briefly discussed in other blogs). This is buying products that you KNOW are not good for you and which may not even taste as good as healthier alternatives -- but if convinced that it is the popular thing to eat (in the popular quantities) then you may choose to buy and consume it.

The "ugliest" part of commercially processed foods is the addictive parts. Processed sugar (including High Fructose Corn Syrup) is an addictive and unhealthy substance. It can be found in many processed foods including canned vegetables. Check the ingredient list! Salt is a flavor enhancer but, for some people, it can be hard on the body in excessive amounts. Governmental control agencies have stopped the addition of such substances as cocaine, coca leaves, and other known addictive chemicals (yes, they did use such at one time) into foods but there isn't enough oversight to make sure that other newer chemicals do not have addictive or destructive effects.

The bottom line is that processing food is a long, and reasonable, thing. Local processing is probably always healthier than commercial processing but there are reasons why people choose to buy commercially processed foods --

187

and why they take more and more shelf space in the supermarkets and grocery stores.

Do you make use of processed foods? What kinds do you purchase? Do you check the ingredient lists when you make choices? Does it matter to you if you are preparing for one or two versus preparing for a large family gathering?

64.   A Body for All Diets –
8/29/2015

The human body is a marvelous set of systems. Eating and processing food seems relatively simple but, as seen in previous blogs about nutrition and inner bacterial colonies, it is not really that straight-forward. Nevertheless, our bodies can definitely be considered omnivorous -- able to eat anything that is not poisonous (and a limited amount of some foods, such as alcohol, that are poisonous).

Our bodies need certain "building blocks". These include calories (for basic energy), amino acids, vitamins, minerals, and basic nutritional blocks (such as fats/oils, carbohydrates, general proteins, and so forth). Since I have already spent a number of blogs on this subject, it does not make sense to go into further detail at the time. The point is, these building blocks are required for a healthy body -- but the form in which we take these building blocks in can vary a lot.

For the sake of argument and categorization, let's say that there are four types of diets available. The first is carnivorous -- eating only meat. The second is omnivorous -- eating foods that come from other animals as well as plants and items that are difficult to classify. The third is vegetarian -- eating only foods that do not require animals to be killed. The fourth is vegan -- eating only foods that are considered to come from the plant side of biology.

As mentioned earlier, our bodies are designed to be omnivorous. Thus, this can be thought to be the "easiest" type of diet. Eating a mixture of meats, plants, and animal-derived foods gives the easiest access to our nutritional building blocks. This does not mean that it immediately gives rise to a healthy diet. For all diets, it is still needed to make sure that quantities of calories and the balancing of other building blocks is kept in mind. Most people in the world are omnivores.

One of my sons calls himself a "carnivore" -- but he still eats pasta (and cheesecake) and other items from the rest of the planet's living matter. A true carnivore eats only meat (some large carnivores eat a small amount of grasses for fiber and minerals but not for calories). For humans, this only occurs naturally when living in an area where plant food is very scarce -- such as the Arctic regions. The Inuit eat an almost exclusively carnivorous diet and, in general, stay quite healthy. Upon study, it has been found that this relies on two things. In order to get a full set of nutritional building blocks, they must eat most of the animals including bones (or bone marrow), internal organs, and other parts that many humans avoid. It is also important that some of the animal bodies be eaten without cooking as cooking can destroy many vitamins and amino acids.

I always thought that vegetarian meant eating only non-animal-based food. It turns out that the original definition of vegetarian meant to eat a well-rounded diet -- to curb the excesses of the diets that were eaten by those who had access to many types of foods. However, in the modern era, vegetarian is considered to be not eating animal tissue. Animal-derived foods, such as eggs and dairy products, are also allowed. A vegetarian diet can be very healthy -- especially if animal-derived foods are kept to a minimum -- but it is very important to watch the needed building blocks carefully.

Finally, there are vegans. The word vegan was derived from the beginning and end of the word VEGetariAN. No animal-derived products are allowed. Since this means eating exclusively from the bottom of the food pyramid, it is difficult to find an overweight vegan. It is also relatively easy to become unhealthy if the diet is not watched carefully. Traditional vegetarian cultures and religions work with this and adopt a cuisine and menu that naturally balances out the needs of the body. A self-imposed vegetarian must devise this for themselves. Luckily, it is much easier to do

190

that with an abundance of literature and products in our global economy.

Which diet is "best"? Telling an Inuit to become a vegetarian would be telling them to starve. If the entire world's population decided to become strictly carnivorous, then probably 90% (or more) of the world would starve as there is not enough animal substance to feed everyone in the world. There are also hybrid eating philosophies such as "vegetarian with fish". Being vegan is the easiest burden on the planet since it uses foods from the lowest levels of the food pyramid. However, this is also not possible for all people, without a global economy, because some areas of the world do not have enough human-edible plant-based food on which to survive (but other animals CAN eat those plants and provide a higher-level food for humans).

In a global economy, we have choices. Foods can be grown in one place and eaten in another although this places a burden on transportation, fuel, and storage costs onto the planet. Locally produced foods are a lighter burden but is not possible everywhere. Many people choose vegetarian (or vegan) diets based on their individual morals that determine that unnecessarily eating animals is wrong. Vegetarian/vegan eating is also less expensive but requires time for balancing and preparation.

Our bodies may allow us many choices in diets but the choices are still individual for people. What choices have you made and why?

65. What is Critical About Mass: Having Enough to Sustain a Process – 10/3/2015

There is a phrase used from time to time -- **"critical mass"**. This phrase was initially used in regards to the manufacture of nuclear weapons. In order for a nuclear device to be able to start a fission (splitting atoms) reaction, there must be enough material that each splitting of an atom creates enough energy to allow for the splitting of additional (at least one more) atoms. So, the critical mass is enough radioactive material to allow for the process to continue once it is started.

This concept is central to the creation of weapons because a bomb is created such that two, or more, amounts that are <u>less than</u> critical mass are kept close together -- when it is desired to detonate, the smaller amounts are pushed together to create the critical mass and the reaction can take place. A bomb is "clean" if it has the opportunity to split most of its radioactive material before spreading apart and "dirty" if it ends up spreading the radioactive material (not fissioned) into the surrounding area.

This concept is also important with the design of nuclear power plants -- rather what NOT to do when designing a plant. A properly designed plant will maintain control of how much radioactive material is allowed to be near each other (with the addition of materials called "damping" rods which absorb excess energy). This is important not only for safety but also for the economic operation of the plant. If a plant is not designed this way then it is just a one-time-use bomb.

Thus, nuclear power plants (though there may be other dangers) are almost impossible to cause nuclear explosions.

One more term that is very important to this topic is the "**tipping point**". This is the very small range that exists between NOT having a critical mass and the amount reaching a critical mass. Just a bit more and it becomes critical. Remove just a bit and it becomes inoperative. Before the tipping point is reached, the process requires continued external energy to maintain progress towards the critical mass.

All that is just a preamble to this blog [smile]. The concept of critical mass enters into many areas of our society -- political, economic, sociological, and so forth.

One area of present interest and, which is entering the region of the tipping point, is that of electric cars. Electric cars have been around for quite a while (according to the Net, 1834). However, we do not see electric cars everywhere -- it is mostly internal combustion (gas or diesel) engine cars. Cars, by definition, are used for movement. This means that the source of their energy must be carried along with them. In the case of gasoline engine cars, this means tanks of gas. In the case of electric cars (actually, electric engine powered cars), this means batteries (or an awfully long electrical cord [smile]).

Batteries have traditionally not been very powerful or very efficient. This continues to change and, although not specifically a critical mass, they are now reaching the point of efficiency to be able to be used more practically. (Note that the same idea is involved with the efficiency of solar cells for solar power.)

Even with more efficient batteries, they still must be charged on a regular basis <u>and</u> the means to charge them must be close enough together (driving range) such that an electric car can go from one charging station to another. Additionally, the time needed to charge the battery must be relatively short or timed such that the charging can reliably be done at "non-use" times (such as night).

So, the technology needed to have practical electrical cars on the roads requires three things: sufficiently efficient batteries to provide workable range, charging stations within that range, and technology sufficient to compete with other alternatives. This is now beginning to happen. We are close to the tipping point. The external energy causing us to reach this point has come from the dedication of various people who want the end result. Note that the same process happened to make the internal combustion car practical in the early 1900s.

Another area of critical mass is concerned with political, or social, matters. Let us take the matter of the ability to vote, within the United States, for women. Within the democratic process, one group cannot grant themselves additional authority, privileges, or rights. They must be granted such by the people who already have that power. This means a process of change of thoughts and attitudes. The energy to achieve that came from dedicated people who worked towards that goal. They achieved critical mass when enough of the existing authorized voters were convinced that women should be granted the right to vote.

There are many areas where the ideas of a tipping point, and critical mass, are important. They are both involved in areas of change. The change may be a chemical, or physical, process. The change may be a social process. The change may be a political process. But they each have stages and move from one to another by approaching the tipping point, reaching critical mass, and effecting the change.

What areas of critical mass do you see approaching and which ones do you see from the past?

66.   Entropy: or Why is my House a Mess? – 11/21/2015

Physics is basically a bunch of descriptions (called "laws", "theories", "postulates", etc.) that allow us to predict how energy and matter (something that can be touched by other matter -- like a chair or oxygen or water) will be affected by other matter or energy that is applied to the original system. Note that, with special relativity (the old famous Einstein equation of $E = mc^{**}2$) it is recognized that energy can be converted into matter and matter can release energy if broken into less complex units. (As usual, it is easier to break something down than to put it together.)

One concept is called **entropy**. This is a measurement of disorder. It applies to the melting of an ice cube. It applies to the complexity of atoms -- with entropy increasing when fission occurs breaking a uranium atom into smaller atoms. It can also be used for everyday things -- such as a tidy room becoming messy. In each case, the situation of order moving towards disorder is called entropy.

Local entropy can be reorganized. This often requires putting in energy -- but not always. The free flowing molecules of liquid water are less organized than that of ice which has a static configuration and form. In order to create ice from water, it is necessary to remove heat (energy). Because there are different ways that order is created from disorder (or disorder from order), there are considered to be different types of entropy.

Notice that I say <u>local</u> entropy. According to the Second Law of Thermodynamics, the overall process of entropy cannot be reversed. If you leave something alone, it will eventually break down. In a house, everything will get covered in dust. A moving part will wear and break. Radioactive particles will continue to convert into non-radioactive masses.

So, whether one believes in creationism or the big bang theory, everything started in chaos, order was imposed, and everything keeps trying to make its way back to chaos. We can clean the dishes but they'll get dirty again (even if we don't use them, dust will settle).

Next time you look into your house and say "didn't I just clean this up the other day?" just tell yourself -- "Oh yes, entropy".

67.   History Isn't What it Used to Be – It Never Was – 12/19/2015

History is a tale of what has happened. But is it really what happened? How do we know? What is the evidence? If we have records of what has happened what about records of what did not happen?

In the Harry Potter series, the instructor of History is dry and boring and most students in the class don't pay a lot of attention -- until something from out of history seems to pop up once again in the present. This happens with the Chamber of Secrets. But really, the same thing happens to people all the time -- they lose track of what has happened in the past until it starts to happen again in the present and then SOME people recall the events in the past. There is the notion of people remembering the past in advance of events occurring -- an attempt to prevent old mistakes from happening again. But this rarely happens. Fashions cycle. Historical events cycle. Political environment cycles.

One thing that does bother many people is an active attempt to hide history of which we no longer find acceptable. School systems start to change books to remove the realities of human slavery in the past (and deny the occurrence of present day slavery). Videos are actively altered to sway opinion and emotions. Some people try to deny past atrocities in spite of hundreds, thousands, or even millions of witnesses. Books are edited to make the language of the times in which they were written change to reflect the attitudes of the present. There have been many stories and novels written about such attempts -- often by governments but sometimes by active minorities. But it is very likely there have already been changes actively made that no one influential has noticed.

Another area in which history does not match what happened in the past is something that might be a coincidence (and might not) which is that (in English) history is very similar to "his story". And it is true that history is predominantly a story of males and male events. Possibly this is because patriarchal (run by males) societies have dominated the past. Perhaps it is because males have been more likely to learn to read and write. It is certainly true that many inventions and creations have happened by women and been uncredited. Check out books such as "Mothers of Invention" (by Ethlie Ann Vare and Greg Ptacek) and "Underside of History" (by Elise Boulding).

Another problem with resolving the accuracy of history is that of translation of primary documents -- those records that were produced at the time of the events. Who wrote those documents -- probably the "winners" of conflicts. What happened from the point of view of the losers? Decimation of the First Nations in the American continents or the subjugation of the Ainu in the Japanese islands or the aborigines in Australia all have forcibly cut out a good section of the ability to balance the events of history.

Even recognizing that only the "winners" and the powerful usually have their documents preserved -- language is not independent of the context of the times. If I write a sentence in English today then that very same set of words is unlikely to have had the same meaning 500 years ago and is unlikely to have the same meaning 500 years in the future. It is horrendously difficult to be able to know what was meant within a language within a context that is no longer present. Books that are the bases of religion often expose this fragility.

In my own family, I recognize how much knowledge has been lost. Where did my mother meet my father? Where was their first date? (I don't know.) What was the occupation of my multi-maternal great-great-great grandfather? What color did he like best? Who fired the first bullet between the Hatfields and the McCoys? Certainly these may be "unimportant" pieces of information but there is no longer a way to know -- the knowledge has been lost.

So, our ability to know what has happened in the past can only be an approximation. Sometimes things are changed deliberately. Often the documents that are saved reflect only a narrow set of perspectives. The understanding of the meaning of documents and language change through the years. And, in general, much is lost because more was lost from knowledge than was recorded. So, if I read an article in a newspaper or book about a current event and they refer to historical aspects -- I can read it and recognize that that is how people think about the history. It may, or may not, be a reflection of what really happened.

But that is not an excuse to not try to do better in the future.

68.   The Houseboat Philosophy :
The Four "R"s – 12/31/2015

Once upon a time, I lived in Seattle -- loved it there because it was an environment in which I could walk, or take public transportation, to 80% of the places that I wanted to go (exceptions being grocery shopping -- mainly out of laziness). A pedestrian-friendly city gives a natural interest in adjusting one's life, in general, to the environment and trying to live within the environment rather than take it over.

My wife-to-be was taking courses at the local University of Washington (Udub) and one of her teachers invited us to a lunch on her houseboat that was docked on Lake Union. While there, and conversing on various topics, she introduced the "houseboat philosophy". A houseboat has a certain capacity -- it weighs a certain amount and it can support a certain amount without sinking. So, unless you want to start swimming for shore during the next rainstorm (when the lake becomes definitely non-calm) you recognize that clutter is not something that you can tolerate.

When something comes onto the houseboat, something has to go off. In a way, this is parallel to time organization of "most important now". What do I most want to have around me? If I bring something aboard, what do I want to take off and what will I do with it? Of course, there are a certain number of things that are in constant transit -- like food. Most things, however, including a stockpiling of the pantry must be considered to be an added weight.

What to do with what one takes off leads directly to the three "R"s of Recycling. Reduce. Reuse. Recycle. Reducing is associated with packaging material and extra things that are not directly part of the item which you acquire. Reuse can be a matter of giving used items to charity to be re-sold or to become part of the lives of other people who need them. It can also be used to indicate a refilling of containers and such. Finally, recycling allows the materials to be incorporated in new products.

The Houseboat philosophy adds one additional "R" -- Rethink. This is a preparatory thought before acquiring something. Is this something that will add to the quality of my life? Is this something that is more important than something else that I presently have around me? This isn't something that happens a lot in the current consumer-directed economy and society in which we live. (It DOES start moving that direction when one starts going towards the final direction of life -- a desire to eliminate the unneeded before someone else has to take care of it.)

All together, the four "R"s are really the Houseboat Philosophy. While, for most of us, our houses are not likely to sink if we get one too many things into the house, clutter still affects our lives. How many times do you have to look for something you have mislaid? Is it easier to find one item out of a thousand that you have around -- or is it easier to find one item out of a hundred? How about cleaning? Is it easier to clean around 20 pieces of furniture or is it easier to clean around six?

We are presently reducing in our household. We have a lot of books. Each book includes memories -- where did we get it, what thoughts arose when we read it, what other people were around when we read it, what was our mood before and after reading it. It is hard to remember that, even if the book is no longer around, the memories can still be present. If you have a box of things that hasn't been opened in ten years then how important are the things in that box?

What are the important items to keep on your houseboat?

69.   Driving and Physics: The Laws of the Road That Cannot be Broken – 1/9/2016

There are two sets of laws involved with driving a car. The first set is made from the human-made laws which are a set of etiquette laws of how people can share the roads and passageways while driving 700+ kilogram vehicles. The other, underlying, laws are associated with the laws of physics. The first set can make interacting with other vehicles, and their drivers, more predictable -- and generate income for the various cities in which you drive. The second set -- unlike the situation for the Coyote or Bugs Bunny in Looney Tunes -- cannot be broken and determine what will happen when your car interacts with other physical objects.

The first topic that I'll mention is that of **relative velocities**. If you are driving 30 mph (or 50 kph, if you prefer) and you run into a wall, there is 30 mph (squared times half your mass) amount of force with which you hit the wall. This is called kinetic energy. If you run into something coming towards you (such as another car) at 30 mph, the amount of energy is quadrupled and it is as if you hit a wall at 60 mph. (Note that hitting a wall at 60 mph does more than four times the amount of damage as 30 mph.)

If two cars are going in the same direction then the collision energy is <u>subtracted</u>. If you are driving at 60 mph along a road and someone driving at 65 mph bumps into you then it is the same as if they drove into a wall at 5 mph. Not very noticeable unless that bump makes you lose control and you use your 60 mph kinetic energy to run into a tree.

This is the principle behind merging. The idea is that you drive your car such that you are driving at approximately the same speed as other cars by the time you leave the on-ramp to the highway. You speed up a little and safely merge ahead of a car. You slow down a little and safely merge behind a car (preferable). If you are going 30 mph while all the other cars are going 60 mph it makes it much more difficult for you, and for all the other cars, to merge safely.

This leads to the next topic -- "tailgating". This is where you are traveling at a speed such that it is not possible to stop without colliding into the car ahead of you if they abruptly stop. My old traffic books indicated one car length per 10 mph -- so, around 90 feet for travelling at 60 mph. I routinely see people driving with a single car length between cars while driving 60 mph. This situation is very dangerous for two reasons -- if the car ahead abruptly stops, you have converted a 5 mph bump into a 60 mph crash into a wall. The second reason is that it makes it very difficult to merge. The merging car cannot safely go ahead, or behind, other cars on the highway if there is no room.

Tailgating is directly involved with another law of Physics called **inertia**. This is Isaac Newton's "First Law of Motion". It says that if something is at rest it will want to stay at rest (difficult to start moving) and if it is in motion it will be difficult to stop and will continue at the same speed and direction unless outside forces change it.

So, let's apply this law of inertia to tailgating. Stopping distance involves the factors of "reaction time" and physical stopping time. Reaction time varies between people. In general, women have better reaction times than men and, in general, younger people have better reaction times than older people. However, reaction time can never be zero as it takes time for the outside signal to reach your eye (you see a brake light) and your brain to process the signal to start a reaction (stepping on the brake).

This is why it is impossible to stop if you are tailgating someone and they abruptly stop no matter how good are your reflexes. IF you both started stopping at exactly the same time AND both had the same tires and same braking systems on your cars THEN tailgating could be safe. Those factors are not true.

This leads into the final topic for this blog. Driving in weather. Weather affects a lot of factors. It can make visibility easier or harder -- which affects reaction time. It can make the traction of the tires on the road better or worse -- which affects physical stopping time. Results from relative velocities remain the same but the outside aspects which make a difference in how, when, and why have changed.

So, if you drive on a snowy, icy road you have increased reaction time, physical stopping time and the principles of inertia make it more difficult for you to either turn or stop without the car wanting to continue in the same direction and speed.

In the end, the laws of physics will exist and they won't change so merge properly, don't tailgate, and allow for changes in road conditions for judging safe driving.

[formula for kinetic energy corrected due to the sharp eye of Dave Albay-Yenney]

70.   Malleable Memory: What You Remember isn't Necessarily what Occurred – 3/19/2016

Memories are that special something that make us who we are. Twins may have the same genetics but, even if raised in the same environment, they cannot have the same memories. You would think that, if this is the case, memory must be the most reliable of attributes of the human condition.

But this is NOT the case. Memories can be formed in ways that they are inaccurate at the very beginning -- and change over the years to fit in better with other attitudes and stories concerning the subject matter.

Joseph Campbell related a local tale of a village that had a mischievous god who visited. There were two rice fields on different sides of a small dirt road. This god put on a big woven hat -- one side was bright red and the other side was bright blue. He put it on his head such that the people working in the field on the right saw the red and the people working in the field on the left saw the blue. He reached the end of the road and turned around -- but, while turning around, he also reversed his hat such that now (going the opposite direction) the people on the right saw the blue side and the people on the left saw the red. Thus, for both walks down the road, people in the field saw the same color.

When the villagers finished their work for the day, they met in the village and talked about this strange man who walked through the fields wearing a bright red hat. "NO" said others he was wearing a bright blue hat. So, they argued and fought and the god laughed. This is one of the first ways that memory is shaky -- no one can observe everything that there is to be observed and different people will observe different things.

A second area that changes the formation of memories is that of expectations. These expectations are based on personal histories, biases, and even current events. During a classroom experiment, two people (without forewarning of the class) entered a classroom -- male and female and of different ethnic backgrounds. They loudly started to argue, fight, and then leave the room. After they left, within five minutes of their arrival and before the students could discuss among themselves, the instructor had the students write down an account of what happened.

When the instructor read through the accounts, she found that there were not two accounts that read the same. Some of the things that the actors did were reversed -- things that the female did were recounted as things that the male did and so forth. The interpretation of who did what first and which one was justified in their reaction also changed. In general, if a history book was to be written from these first-person accounts there would have to be one per person.

The third area of moving early memories is that of peer influence. Once a situation is discussed, many people will start changing their memories to that of what the most popular people remember -- or will allow for change based on arguments presented by others of more vocal temperament.

Within a few days, the people no longer remember any different account. These are some of the many aspects of how initial memory can be altered. In the next blog, we will talk about longer-term memory and how it changes with time and can be altered.

71.   A Divergent Possibility: Reuse or Recycle is a Fork in the Road – 4/9/2016

The three "R"s include the areas of Reduction, Reuse, and Recycling (with the addition of the preliminary "Rethink" of the Houseboat philosophy). But how is a decision made whether to resuse an article or to recycle it?

Generally, this is a matter of how easy is it to do one or another. To reuse, it must be able to be used -- it must be in working order. A book is presumably able to be reused easily unless it is damaged. An old sofa might be moderately damaged -- needing a hole to be repaired or a leg put back on. A piece of electronic equipment may require a diagnosis followed by repair. There is a sliding scale of ease of reuse. No work is needed up to potentially quite a bit of work (perhaps as much as creating the item originally).

Sometimes, repair can become part of a hobby or enterprise. People may take great pleasure in the work needed to repair a vintage car -- and the parts and labor can be considered an investment. The same can be true for people who buy a house in need of repair and succeed in renovating it such that it can be lived in, or resold, in a much better condition (and higher price).

But, what about a toaster? There are some "vintage" toasters that can rise in value -- but most of them are "commodities" -- able to be replaced easily or replaced with "improved" characteristics. Is it worthwhile to repair? Is it a part of an internal enjoyment to repair? Possibly for some but for most it is not worthwhile. So, we hit an economic tradeoff which is very dependent on the local cost of merchandise and the local cost of labor.

Let's say that a new working toaster is priced at $30. A used working toaster can bring $5 at a flea market or local swapmeet. If the cost of labor to repair toasters is $20/hour and it takes two hours to repair it, then making a non-working toaster able to be sold would cost $40 and only sell for $5 giving a net loss of $35. Who would decide to do that? Hopefully, the person would find a good place to recycle the toaster.

However, if we are in another place where a new working toaster is priced at $20 and a used working toaster can bring $7, we may have a different situation. What if labor costs are only $2/hour? A non-working toaster can be repaired for $4 and sold for $7 -- giving a total profit of $3. We can see that the local prices and local labor costs make a big difference as to whether something is reused.

Another factor that comes into play for reuse versus recycling is convenience. If I have a paperbook that is not severely damaged I have a choice for reuse by taking it to a used bookstore or recycling it via a curbside pick up. If it takes me 40 minutes (roundtrip) to take that book to the bookstore then recycling becomes an attractive option. If I have a box of books, however, the time needed to take them to the used bookstore becomes less of an overhead. Convenience and net savings come into play -- plus the value one gives to their own time.

As an umbrella over the choices is the priority one gives to the environment. Reduce, Reuse, Recycle are the three rules but they only apply to people who care about the effects of NOT doing these things. Some may care because of the money involved with not doing it -- some may care because of the environmental costs -- and some may not care at all. What is my time worth in choices needed to help maintain a good environment? Do I care about the state of my surroundings when the next generation is growing up? These are underlying questions with individual answers.

.

72.   Solidified Memory: How do we Make Sure We Remember What we Experienced? – 4/30/2016

If memories are so unreliable (and they truly are) then what can we do to make them more reliable? How do we convert them from transient to permanent? Well, on a technical basis, that process is called moving them from short-term to long-term memory. But, as we have discussed, that still doesn't make the memories any more "true" just because we remember them longer.

First off, it is good to acknowledge that almost any "evidence" can now be forged -- with enough money spent, it can be forged to the point that it is impossible to detect that it was not from reality. Technology has moved along a lot since the movie "Rising Sun" was made with Sean Connery. Photos, and video, can be created pixel by pixel. Cursive writing can be reformed from templates into any combination of letters, words, and sentences that may be desired. Audio is still, currently, the hardest to forge as it is a representation of analog (continuously changing) information -- but it is still possible to create something that would be very hard to distinguish from "real".

The only real protection individuals have from data forging is that most of us are NOT "worth the trouble". Even though it is easier and easier, with more easily used tools, to manipulate and fake data it still takes time, knowledge, experience, money (for equipment and software) and (for the second mention) time.

So, how to create a record for oneself that we can use to keep our memories "solid"? We have already inferred a number of methods. Photos. We can take photographs (preferably with notes associated with them). We do have to be careful to not infer about what we do NOT see. If we take that photo of a person in a red hat -- we do not know the color of the other side of the hat. We do not know who hid behind a tree. However, if we are taking these photos for ourselves, we can include notes to help get the information correct.

Audio is a great way to note things down -- but audio of what we experience is more useful than our taking an audio note of what we noticed or experienced. (That is just as easily misled as any other account of events.) If we can record someone saying something then we can refer back to it as what they said. Content, sequencing, and who said what can be noted and solidified.

Video combines the two of image and sound. The more the merrier. We just have to make sure that we recognize that what is not recorded may be as important, or more important, as what IS recorded. Do not take the "evidence" further than what is really there.

A final method of solidifying memory is just that -- physical reconstructions and recordings. Statues and sculptures, writing, models, and so forth can capture events and the emotions of events.

Do these methods make memories more reliable? No, not really -- although they can record details that we can forget ("Did I wear a green shirt that day or the plaid one?"). But they do help them to stop changing.

73.   The Value Equation: A Formula with No Fixed Parameters – 5/7/2016

People are always searching for a "good value". But what does value mean? Value is certainly a subjective matter -- the value for one person will differ from the value for another person. Still, even though it is variable, there are certain things that make up value. I call this "the Value Equation". Real economists may very well have a better formula for this concept -- but I like mine for general simplicity and use.

My Value Equation is Quality * Quantity * (1 / Cost) = Value. I would love to have Cost not be used as a reciprocal (divided into 1) but that is really how it relates to value. The lower the Cost the greater the Value -- a reciprocal situation. There is one other factor that indirectly affects the Value Equation. That is Affordability. It interacts with Cost in some manner but I don't know how to directly put it into the equation. Let's just say that if you have more money available then cost becomes less important.

Quality is the most subjective part of the equation. If one person really loves something then their perception of quality increases. In a similar way, if another person really hates something then the quality becomes less. It is even possible for it to become negative. A negative quality would indicate that the thing being evaluated goes against moral values. If you hate internal combustion engines then a more powerful engine has less quality for you because it uses more gasoline and probably emits more pollution. If you don't mind internal combustion engines then a more powerful engine has things about it that you love and will increase the quality for you.

Quantity is the only part of the equation that is mostly fixed. I say "mostly" because it is not always true that "more is better' for some people. A huge drink ("big gulp") sounds great until one evaluates the health consequences of drinking too much sugar or artificial sweeteners. A huge sandwich that is more than one should eat either becomes "wasted" (thrown out) or "waisted" (accumulated as fat into your body).

Even cost is a variable factor. First, the price that is charged is not fixed. Often, wealthy people (or celebrities) are charged less for things because they can "take it or leave it" and because their possession and use of it provide a return advertising value for the supplier. Second, in most instances cost does not reflect "total cost". Total cost is the price of all events that exist from harvesting the raw materials to manufacture to distribution to reclamation of the object and the environment from which the raw materials were harvested. Total cost is rarely used -- a good portion of the cost is absorbed by the general population and subsidized by the taxpayers.

Even with the subjective variability of the parts of the equation, it is still easy to see how it is used. Something that is of high quality, in the desired quantity, at a low cost will give the greatest value. If the quality goes down, the value goes down. If the cost goes up the value decreases (but may not be so important if it is very affordable).

How would you define quality? Do you see a limit on quantity that provides value or is "the more the merrier"? Do you consciously take into account "total cost" when you buy something. Do you "waste" or "waist" or do you try to always get just the correct quantity?

Although the Value Equation can be used as a framework -- the final answer is still up to you.

74.   Troll Attack: the New Playground Bully of the Internet – 6/11/2016

The more that a person uses social media, the more likely it is that you will be the subject of a troll attack. Like most situations of being attacked, the most direct way of avoiding such an attack is to not be present. Within social media, this means do not post -- do not show any indication that you exist. For some people, this works pretty well -- they "lurk" in the background and notice what is happening in other people's lives and what are the topics that others are interested in discussing.

There is nothing wrong with passive use of social media -- unless there is something important going on about which you want to give your opinion. Or some event is occurring of which you are particularly proud -- maybe an anniversary or an award that your child has won. Or you keep encountering situations where you say to yourself -- "yes, that is possibly true -- but this other point, which I think is much more important, keeps being missed". Perhaps they keep talking about point B and they don't talk about point A which is related but not the same.

Pride, knowledge, a desire to contribute all may add to the reasons why you may want to actively participate. But, once you are visible, you can (and probably will) be attacked.

You may have heard about flame wars -- which are related to troll attacks. In both cases, the person attacking is primarily making personal attacks -- the messages have little, if any, relation to what was actually said. Many times, there is no way to tell from the attack whether the original message was read at all. However, the sneakiest (and most hurtful) troll attacks occur when they DO mention something that is somewhat relevant to what you have said -- because it can seem like the personal attack is based on some part of your message. It isn't.

Why would a person conduct a troll attack? It truly is the same situation as a "playground" (or workplace, or other locale) bully. First, they are angry. You don't know why but they are angry. Perhaps they lost their job or were refused a promotion in a job or their child was struck by an unlicensed (and uninsured) driver or they dropped a bowling ball on their foot or they were just screamed at by their mother or cousin. You just plain do not know. The only thing you know is that they are angry.

Second, they want to hurt others. Why do they want to hurt others? Why do they want to hurt you? Unfortunately, many people seem to feel that if others are feeling badly, or are being treated badly, then they -- in comparison to the others -- are doing better. This happens with various prejudices as well as in the arena of bullying.

Why are they attacking you? Well, it is possible that your message mentioned a keyword -- or "hot button" -- something about which they have an entire set of preset ideas and emotions. Am I saying that they ARE addressing your message? No -- not at all. They are still creating a personal attack based on anger. They are not listening (reading), analyzing, researching, thinking, or anything that really is concerned with the topic. They are reacting in anger because they are angry.

Why else might they choose you to attack? It might be because your message shows some indication that you may be vulnerable to attack -- it indicates insecurity or that you are hurting -- and there are also "-ist" attacks based on peripheral things like religion or pigmentation or ethnic origin or such. They attack because you seem to be a person who can be attacked.

What do you do if you are attacked? Personally, I suggest to do nothing. Like other bullies, if they do not get anything back to feed their anger, they will choose a different target. (It is conceivable, but unlikely, that they will even reconsider what they have said and try a different response -- at which point in time it is up to you whether to respond.) It may be useful for you to create a reply -- but never, never send it -- in order to allow emotional responses to find a form.

It is possible that others will come to your defense -- which may have some effectiveness unless they also start attacking from anger rather than addressing the contents of the messages.

Disagreements occur in life -- and they occur on the Internet. When the contents of messages are addressed, researched, explained, expanded upon, and treated as a bit of information within a much larger pool of information, those disagreements can lead to growth, change, and continued exchanges of ideas. When people attack the person who writes the message then there is no route to constructive response. It is an emotion not a thought and they may have reasons to be angry -- but they do not have the right to address that anger upon you.

How do you deal with flames and troll attacks? Do you have any methods that work to decrease them?

75.　Social Media: Still No Such Thing as a Free Lunch -6/18/2016

I recently had a friend, who also uses one of the same Social Media sources that I do, complain about the way their contacts list was being used to send out advertisements to her friends under their name. For them, this was an item that made them consider dropping use of that Social Media source (I am deliberately not naming it because the problem is not really specific to that particular source -- call it YYY.) I responded to their message with a brief note about how all of the Social Media sources had to find ways to fund themselves and that if they chose to drop usage that was certainly their right but to recognize that the source had to be able to fund themselves.

During the past 50 years, we have had a true technical revolution -- meaning that the ways that things interact have dramatically shifted. All change causes discomfort and the need to adapt new methods to work with them. However, as I have talked about in some earlier blogs, money -- which is the representation of labor and other resources -- still needs to be able to be moved around so that people can pay for their needs to live.

It may be difficult for many people to remember so far back but, once upon a time, everything was paid for in cash of the local economy -- or, possibly, private representations of cash such as personal checks or money orders/traveler's checks. My younger children have never written a check -- and it is quite possible they never will (they also may not learn cursive handwriting to be able to sign a check or contracts -- but that is a different story). The first credit card (or what we would call a credit card) was invented around 1950. For the first couple of decades, a credit card was used more as a guarantee against payment with the card's numbers (sometimes imprinted from a raised surface) associated with an account which was then printed with the purchase/fee amount and sent to the local bank or credit repository. The money was then authorized to be given to the merchant and a bill was later created for the person using the credit card. It was not until the 1980s that the landline phone system began to be used to connect directly to the credit card issuer's accounting system -- an "electronic" credit card. Of late, it is becoming popular to embed "smart chips" to increase security.

With use of electronic credit cards, the user, the merchants, and the credit system became part of the "big data" pool of information. Privacy was greatly diminished -- laws were created to help protect privacy but certain information could now be accessed unless directly forbidden. If you buy a specific product at a specific store, you may get (in the mail -- or via email) a coupon for a competing product at a competing store (or the same product but at a different store). They know WHAT you get and WHERE get it -- and the purchase is specifically connected to YOU. The advent of even more abstract methods to transfer money such as Paypal means that all info within the capital stream can be matched against each other.

So, with that general background, here comes the Internet. Others are better qualified to talk about how it evolved than I am -- but it basically started as a network interconnected by the Defense Department to help its contractors better communicate with each other. That expanded into a general connection of universities, colleges, and scientists which then expanded into connections between businesses as well as all the former connections and then the leap occurred for virtually everyone to connect with everyone else. This global interconnection used to be via voice phone, or physical letters and telegrams, or long personal trips.

The old, physical, methods had a set of costs to provide services and a set of fees and charges to make sure all the people involved could continue to provide the services. The old pre-Internet was paid for by the Defense Department, and then divided between the Defense Department and the various businesses and then private companies started to be formed which helped with interconnection for a fee -- with transmission fees to the private companies paid directly by the end user. That is the way that the fee structure is largely set up now. People pay for connection (cable, DSL, analog phone, broadband ethernet, whatever) and usually have an Internet Service Provider (ISP) -- many times these are now provided by the same company. In some manner, the full amount of fees/charges must pay for the needs of all the people/resources needed to provide the service.

We (finally, you say) now get to the Social Media. Social Media is a destination -- just like going online to shop is a destination -- or online to get information is a destination. Each destination has an interest in having you go there. But every destination has its own costs needed to provide the services that are attracting you to go there. Online shopping sites are a straight-forward equivalent of a "brick and mortar" store. Their costs are paid for via the profits on items that they sell. Online information access is usually paid for by the people who want you to have the information -- tourist destinations or government entities (taxpayer funded) or whatever. Private information destinations may be paid for by advertisements which exist to redirect you to businesses which have use an online shop finance model.

But what about the Social Media? Every destination has to get you to decide to visit. Many of the major Social Media (and major "search" groups) decided that the services would be "free" to the user. In other words, people could work with the destination's services without paying any additional money directly. A "free" site can then entice people to come there by providing the services that they want to use without having to precisely decide on a fee structure for the services (which, if they give many types of services, could end up being very complicated). BUT, the Social Media still has people and other resources that are needed to provide the service -- and these people have their own needs to be able to live. So, every destination -- including Social Media -- must eventually bring in money to pay for those services.

Once a service is provided as "free" it is very difficult to start charging up-front fees without having a mass departure of people from using the service. So, the fees must be charged in a manner that is "optional" -- you are not required to pay/use them in order to get the general services -- but enough people are expected to want to use them such that the money brought in is enough to pay for costs. The first, easiest, avenue is advertisements -- this has a long-time association with use of services and people expect it (even if the service initially starts with no advertisement). It is even a way to get initially "free" services changed to a fee-service without getting rid of users -- ("free" with advertisements but, if you don't want ads, you can pay a fee to get rid of them). The next step is to provide access to other services which, once again, may be "free" but have added inclusions that do cost money (for example games that allow you to purchase "extras"). A following step is to "sell the client list" to other fee-based advertisers (such as my friend complained about).

There are various methods used to bring in the revenue needed to provide the services and resources -- some are very ingenious. The goal is to make you WANT to pay -- something that you have been persuaded that you NEED -- without ever making you doubt the reasons for which you initially went to the destination. It is a "tightrope" for some companies and they often sway back and forth between not making enough money to starting to lose people because they are unhappy.

What other methods of bringing in money do you encounter? How do you feel about them?

76.   Magic Numbers: Society and What is "Normal" – 8/7/2016

I was involved in a thread once upon a time and got trolled (expected if you talk about anything of significance -- and sometimes even if you are just talking about the weather). This was a thread talking about one of society's "magic numbers".

These are part of a group of numbers which we use based on statistical information. As Mark Twain once said (he said that the British Prime Minister Disraeli said it first): "There are three kinds of lies: lies, damned lies, and **statistics**."

For example, in current society a number called the Body Mass Index (BMI) is used to determine whether you are overweight, underweight, or in a "good range". This number is a ratio of height to weight and provides reasonable results for 80 to 85% of the population. For the other 15% to 20% of the population, the number just doesn't work well. If you are an athlete (or work out a lot and have more muscle) it doesn't work well. If you are a total couch potato with almost no muscle it will actually give you a better result than you "deserve". If you truly do have "thick bones" you will be at a disadvantage. This doesn't matter that much except if people are basing other things on that number -- insurance companies and computerized social services for example.

Why do we use this "magic number"? It is quick. It is easy. It is cheap compared to other, more accurate methods of body fat percentage calculations. However, even if one used a more accurate measure than that of the BMI, it would STILL not be accurate for everyone because there is no "one body shape fits all". For some, to hit that "ideal", one part of their body would have to be UNDER-weight in order for the average to be correct.

There are other "magic numbers" used. The "age of consent" is a magic number which, in more technical society, indicates that a child has become an adult. They are able to sign contracts, do things without permission from parents or guardians, get married and so forth. This magic number ranges from 13 in Japan to 21 in Bahrain. In some non-technical societies, it depends on the age when menarche sets in for women and "rites of passage" for men.

Many people get the direction of "age of consent" backwards -- thinking that a higher age of consent protects the child more from society. In reality, the family, church, or government usually have many options to do what they want with children at the age they think is right. The "age of consent" is what gives people the right to control, for themselves, decisions that affect their lives.

These numbers are set for two principal reasons. One is to prevent abuse of others by pushing them into activities (such as marriage or other things) before they can really make good decisions for themselves. The other is to determine an "age of emotional and mental maturity" which is taken to be an indication that they can make good decisions.

This type of magic number is determined almost solely by societal norms. It is usually lower in agricultural societies and higher in societies that require a longer period of education and social adjustment. It is also higher in societies where familial, and religious, control of women is greater. But, as is true with the BMI, it is (at best) a statistical reflection. Some will not be ready at age 30 to properly make decisions for themselves. Some might be ready at a very young age. There have been no psychological studies and are unlikely to be such.

Even the "age of consent" is not for everything. There are separate "magic numbers" for voting, being drafted for war, being able to drive, being able to work full-time, for purchasing and using legal drugs, and so forth. In many instances, the society would **like** to prohibit the activity but do not have sufficient backing from the populace to do such. Therefore, they set an age which most of society agrees is appropriate.

In all of these areas (and more), the "magic number" is sometimes determined by a statistical averaging and sometimes determined by societal norms. It is rare that the number is backed up by thorough, and consistent, studies -- which is why it is "magic".

What other "magic numbers" are you aware of within society? Do you know of any that have a researched background reason? Others that do not have such a background?

77.   Labels and Marketing: What's in a Name? – 8/16/2016

This blog is about marketing -- and marketing is about politics (and politics is about marketing). Most people think of politics as about politicians and other elected officials. Most people think of marketing as being for/against commercial products and other types of physical objects that can be transferred. Yet, we talk about marketing a candidate and politics are basically a matter of forming/creating, and expressing, views on various subjects. They are inseparable.

So, this is a warning. If you don't want to read a blog about politics then don't read this blog. Some examples will be about commercial products and other examples will be about current events and, admittedly subjective, aspects of current society.

Sometimes marketing is used as a short term for marketing media which is about HOW the marketing materials are distributed -- television, pamphlets, advertising within other social media (movies, plays, ...), and so forth. This is different from the marketing content and I won't be talking about marketing media (at least, not in this blog). Marketing is also closely related to sales -- and I won't be talking directly about sales. This blog is about marketing and labels -- the words that contain the concepts that you want to have others absorb.

As mentioned above, marketing is basically a matter of creating/forming views on various subjects. The person, or group, that is doing the marketing will have an object view -- what she, or he, wants you to have as the final thought about the item being marketed. Candidate X is the best. Product Y makes your clothes cleaner. Legislation G will make you safer.

The following bullet items are concerned with categories of methods used to make labels more effective -- more likely to achieve the desires of the people creating the labels. It is meant to include the more important methods but certainly will not contain all of them. Please note that, although examples are taken from the U.S, examples abound from around the world -- in my research, I have found items from Canada, France, the U.K., and Germany.

- Obfuscation -- this is a $20 word for making things less clear. Clarity is sometimes desired within marketing and, at other times, it is strongly not desired. In general, if there is one side then there is the opposite. Gas-conserving versus gas-guzzling. But "gas-guzzling" isn't appealing so that might get ignored and "powerful" would be used instead.

    For a controversial social/religious area, the sides should be "pro-choice" versus "anti-choice" or "anti-life" versus "pro-life". But, even though it is accurate, "anti-choice" isn't something that is marketable. The people opposing the "pro-life" people are strongly supportive of all phases of life, including the post-birth lives of the woman and fully-developed child -- so "anti-life" is not at all accurate.

    In most cases, obfuscation can only occur when the media allow it to happen. Many times, the media will actually assign the labels and they often do so based on "catchiness" rather than attempted accuracy.

- Conciseness versus Self-explanatory and perspective. Shorter descriptions are more easily remembered. "Jingles" are short phrases that are easy to remember (and, often, sung) and get associated with a product. Hashtags are now used to give a short, compact, indication of the subject matter. A difference of perspective can mean that a concise term does not have sufficient information to be self-explanatory.

The term #BlackLivesMatter was created based on the reality that, currently, the lives of black people, poor people, Hispanic people, and First Nation ("Indian", Native, indigenous, ...) people do not matter very much to the U.S. justice system. Their deaths are under-investigated and legal matters are not treated with equal importance. They are also profiled and subjected to laws that are directed specifically towards their communities. They are over-represented in the for-profit prison system.

(What do for-profit systems try to do? Increase demand -- which, in this case, means to increase the numbers of people imprisoned and it is much easier to do that with less-powerful segments of the population.)

People who are not in these groups assume that the groups of people are treated in a similar way to the way that they are treated -- thus they respond that #AllLivesMatter which, unfortunately, is not accurate. It **is** true that #AllLivesSHOULDMatter. If the original hashtag had been #BlackLivesAlsoMatter would that have headed off some of the arguments? A concise label may be accurate but, without appropriate history and perception, may not be sufficiently clear.

- Inferences -- drawing on history and associations. The Patriot Act (and the Patriot missile) have no direct connection to anything "patriotic". By associating the name to the legislation and product, the associations that people have with the word can be connected, in people's thoughts, to the product. Local company names that have their local city, or neighborhood, as part of their name tend to attract more business.

- Avoidance of "hot words" by redirection. Some words are associated with negative things by the majority of a population. In the U.S., such hot words include "socialism" and "welfare". So, if you are going to promote a new taxpayer-funded airport, you do NOT use the word "socialism" (which it is) -- instead you use the less direct "subsidized". If a new factory needs special utility lines (electricity) and water lines that are taxpayer-funded, they are "expanding infrastructure" even if it is specifically for use by the company which does not pay for it. "Welfare" is a bad word when used for helping individuals -- so the word is not used when corporations are subsidized.

- Newspeak -- the way of using language for redirecting thoughts. This is closely related to obfuscation but is more deliberate. Newspeak was described in the book 1984 by George Orwell. A more complete list can be found in the Wikipedia article at https://en.wikipedia.org/wiki/List_of_Newspeak_words.

"Right to Work" laws actually reduce the wages, and freedom, to work for a living. The phrase "Anti-Union" is much more accurate but not as easily sold. The "Internet Freedom" act moves the control of the Internet to large corporations -- thus actually reducing the freedoms of the consumers.

Much of the transfer of wealth from the 99% to the 1% has been done under Newspeak titled Legislation labels. Who has time to read a 2,000 page bill when the extremely inaccurate and misleading title "Help out the Middle Class Act" (not the actual name of a bill, as far as I know) says it all?

Newspeak also deals with the re-writing of history. Don't like what actually happened 50 years ago? Change the history books to indicate something different. Does a book written 100 years ago use language that is no longer acceptable? Rewrite the book to be currently acceptable. This is related to labeling as it is possible, with effective marketing, to change the current meaning of a word. For example, the word "gay" is no longer used in the U.S. to indicate happy and joyous. The label has changed because the underlying meaning has been diverted.

- Irrelevant but popular associations. A carbonated beverage (soda, pop, soda pop, sparkling drink -- regional names) can be marketed as "gluten-free". Carbonated beverages almost never have gluten but, since many people now consider gluten to be bad, a marketing method can include saying what "bad" things are NOT included. "This soda is gluten-free, cyanide-free, lead-free, and contains all the vitamins that start with the letter B".

Labels are a way of marketing products and ideas. As an aspect of language, they make use of -- and form -- the ways that we think about those subjects. If we are aware of what we are doing when we create labels, they can be more effective -- and, if we are aware of what others are doing when they create labels, we can make it less easy to be misled.

Are you aware of other methods that labels can be used, or misused, within the world in which you live? How much do labels affect the way you think about things?

78.   GIGO: Garbage In and Garbage Out – 8/27/2016

In the world of computer science, where a lot of acronyms were created (and still are), there was a term GIGO. GIGO is an acronym for "Garbage In, Garbage Out". This is a shortened version of saying that if what goes into a system (a computer, as one example) is not legitimate then it cannot be processed into an appropriate output. For example, a cruise control system takes into account current gear ratio, axle speed, tachometer reading, and so forth. If the cruise control believes that the car is moving at 100 miles per hour (60 kph) then it will do inappropriate things to the brakes and accelerator if, in reality, the wheels are spinning freely on a patch of ice and the car is not making any forward progress.

If a data entry person (who are almost always touch typists -- not looking at the keys) had their hands at the wrong point on the keyboard as they entered data into a system, then things are not going to go well. You may get a $1,000 refund or you may get a bill for $10,000. In the first example, there is an assumption -- that the speed of the axles is a reliable indication of the speed of the car. In the second example, there is a lack of verification that the correct data are being entered.

Data can also be wrong if the input instrument fails. If your thermostat breaks, then it cannot give proper information back to the processor that controls the oven temperature. You find out when the turkey comes out raw from the oven or the bread starts burning and catching on fire.

Unfortunately, we are not always able to know, or immediately be able to be certain, that a problem has occurred. We may come up with a result that is completely false -- but we do not know that. This is why it is so important to have secondary, or backup, systems and studies to make sure that results are consistent. In the case of scientific studies, one study may be interesting and have results that may entice OTHER researchers to try to duplicate the study or try a different approach, but the results of the one study cannot, and should not, be relied upon on their own.

In the past decade, the term GIGO has expanded a bit. It now is sometimes used for a situation where any type of inappropriate incoming conditions results in bad outgoing conditions. Eating "junk foods" is an example of this in the area of nutrition. If you eat foods that do not supply the appropriate "building blocks" (see my earlier blogs on nutrition) then it is difficult to build, and maintain, a healthy body. Another situation might be a building that is created with bad materials (inferior steel, poorly mixed concrete, lack of specified reinforcements, ...) and it collapses when a problem (perhaps an earthquake for which it was SUPPOSED to have been adequate) occurs.

Another situation is more human-directed. That is falsification of data. In other words, people can (and do) sometimes lie. This may be for many different reasons -- they want different results to be true or they think that other results will bring them more attention (and funding) or they trust that no one will bother to cross-check the results. Two instances (I'm sure there have been more) of this have happened in the past decade -- one dealing with autism and vaccines and the other having to do with climate change. While it is difficult to be certain of the motivations, the results were considerable -- a loss of trust in vaccinations (and a rise in preventable illnesses and deaths) and possible delays in addressing environmental problems.

In the area of politics, of course, this happens all of the time. A politician will say something and reactions and decisions are made based on what they say. If what they say is truthful, then those reactions and decisions have a better chance to be good ones. If what they say is not truthful, then it is unlikely the results will be good ones. Unfortunately, the rules of scientific studies are rarely followed in politics -- the facts may be researched but the results of those fact-verifications are either not seen, ignored, or misbelieved based on what the recipients want to have happen.

A situation that lies between the two is the matter of data gathering for political purposes. Polls are used to reflect, and to influence, people's attitudes. However, a poll may gather data that is not representative and, thus, the results are not accurate. For example, if a poll only calls people on "landline" phones then the people who have only cell (mobile) phones will be excluded. It turns out that certain groups of people are more likely to have only cell phones -- thus the poll will be skewed away from the cell phone group and not be accurate. Or the poll could call people only in the evening hours -- and certain groups not present during the evening hours will not be represented. Or some people will filter out calls based on incoming numbers and so only those who do NOT filter their calls are represented in the poll results. It is more difficult than ever to get an appropriate poll base.

Finally, a situation that combines computer science, politics, polling data, and cross-checking -- voting. There are two types of fraud that can exist in elections. One is voting fraud -- where someone who is not entitled to vote (or who is entitled to vote once but votes more than once) is able to vote. This situation sounds scary but actually does not happen a lot. The other is election fraud. This is where people who should be able to vote are prevented from voting, or their vote is changed such that the person/cause for which they are voting does not get credit (possibly giving it to the opposing situation). This happens much more often than voter fraud and appears to be increasing in volume. Both voter fraud and election fraud are instances of GIGO in the political arena. They are addressed in the same ways -- simplification (fewer systems or people between the voter and the recording of the vote) and cross-checking (paper trails for electronic voting systems, receipts for voting records, duplication of systems and verifying that both results are the same, ...). Unfortunately, people often decide to make things "easy" and "fast" which tends to increase the opportunities for election fraud.

In what areas of life do you see the principle of GIGO operating? How do you, or would you, make sure that the information, on which you make your decisions, is correct?

79.   Jazz and Laughter: Participation Makes the Difference – 10/9/2016

Have you ever been in a room, talking with someone in a quiet corner, when a sudden noise comes from another section of the room? There is another group of people over there and they are very animated and the noise bursts out again. It is someone laughing -- and others in the group are joining them and obviously having an enjoyable time. Yet that noise, even once it is recognized as someone laughing, may not sound that pleasant. Some laughs are called "lilting" or other pleasant references but others are sometimes compared to sounds of other animals or objects in collision.

Whether the sound, at a distance, is pleasant or unpleasant, it can still be considered very appropriate within a group who are all participating in the interaction together. It is this participation that lowers the guards, and criteria, and allows everyone to relax into a mutual experience. If you hear it from a distance -- not as part of the group -- it is a noise. If you are part of the experience it melds into the overall situation.

It is not always necessary to be with the group in order to participate. In the above case, you are unlikely to be able to hear and see what is going on without being among the others. In the case of music, it is often a degree of intensity. In order to immerse in the music, it is necessary to be able to hear it properly. That may mean being in a quiet room, with other quiet people, so that all of the sound can be noticed, listened to, examined, and felt.

On the other hand, it is just as valid of an experience of music to be in the midst of an explosion of sound and people and participate in the emotions of the fellow concertgoers and the movements of the band -- in spite of the fact that it may be so loud that the notes can no longer be distinguished from each other. In the one case, it is the music that is experienced and, in the other, it is the experience that is set to music.

Jazz is an interesting juxtaposition of music and experience. A jazz piece, even when played by the same group of musicians, is not expected to sound the same twice in a row. The variation expands further when it is played by a different gathering. Although a spectator may not be directly singing or playing an instrument they have to be an active participant to fully take part. There are factors of anticipation -- what will happen next -- and surprise -- not expecting what did happen. The music will flex according to the weather and the internal needs of the players and the audience.

There has been much speculation about whether robots could ever "replace" a human. Alan Turing presented what is called the "Turing Test" which says that if you are separated from a computer that can give responses -- so you have no direct knowledge of whether it is a computer or not -- and a human cannot tell whether or not it is a computer giving responses then it "passes". It is truly Artificial Intelligence.

I would submit that an even better test would require the AI to be able to participate within a set of people and know when to laugh.

What activities come to mind when you think of a need to be an active participant.

80.   The Luddite Effect – When the New Does Not Transition the Old – 10/29/2016

In the eighteenth and nineteenth centuries in London, as part of the "Industrial Revolution", a group of workers in the textile industry started gathering together to fight against technical replacements for their labor. Their fear was based in reality. The textile industry in England was a large one within which a considerable portion of the workers earned their living. A mechanized loom might replace the manual efforts of dozens of women and men.

Similar to the situations that often exist today, these people were hard-working and had developed their skills over their lifetimes and, sometimes literally overnight, there was no longer any market for those skills. The response -- a losing battle -- was to destroy machines, make threats to those who were instigating the changes, and disrupt the ability for the new factories to produce. Some historical accounts indicate that the leaders of the workers recognized that there was no way to defeat the change but wanted better leverage to provide retraining and support of the unemployed.

Government response was primarily organized around protecting the new factories, their owners, and products. Severe laws were passed and a number of "show trials" were held with death or penal transportation/exile as potential penalties. These laws, in effect, did succeed in breaking the movement.

Other areas of skilled labor were also displaced within the context of the Industrial Revolution. Although history books usually focus on the improved ability to manufacture goods (and decrease of prices for the average consumer), they do not often indicate the huge labor displacement which was a direct effect of the change.

The Luddites provide a practical history lesson. Change is difficult for societies to adopt and it is particularly hard on those who have invested much time and effort on the old. If change is to happen (and it is difficult to avoid it) then the process of moving away from the old must be kept in mind.

There are a number of changes currently going on in current times. One is semi-involuntary, one is semi-voluntary, and another is fully voluntary.

Climate change is semi-involuntary. This is because it was probably avoidable but made difficult to avoid because of inertia of old methods of business. Although there is still the chance to make the change less severe, it has already made significant changes to the world. The Great Barrier Reef is close-to-death largely because of the increase in global water temperature. The glaciers continue to shrink around the world -- this is especially important in the Asian subcontinent where winter storage of water in snowpacks and glaciers provide water to billions of people. "100-year-floods" and "100-year-storms" are occurring more often as the water temperature rises.

A semi-voluntary area of change is the shift from non-renewable energy sources. Since the change has to be encouraged, and pushed for, it falls into the voluntary category. It is reaching the tipping point where it is almost easier to use new, renewable, energy sources than to keep using the old ones. However, just as happened in the textile industry, it is very important to recognize, and assist, the people and families dedicated to the old energy systems. Solar panel factories located at old coal mines to allow easier transitions?

A full voluntary area of change is the strong push towards greater and greater independent automation. Phones get smaller and more powerful. Robots can take over more manual labor in a programmable fashion (as opposed to dedicated design such as in the textile mills). Innovation and extensive education becomes more and more necessary for general job positions.

Whether voluntary, involuntary, or a mixture of such, change requires preparation and assistance in moving from the old. This is a necessity for the change and, when it is forgotten, much suffering can occur as well as rebellion (isolated or global).

What happens to the old when the new comes? This is an age old question but, with more rapid change comes the need to actively address the needs for migration, retraining, and restructuring.

81. Gambling: A Matter of Risk versus Reward – 12/15/2016

When you hear the word "gambling" you start thinking about casinos, and roulette wheels or maybe hands of poker. But, in real life, gambling is a matter of risk and reward. Crossing the street involves risk and the reward is getting to your destination across the street. Asking someone out for a date involves risk (emotional and, occasionally, physical) in the hopes of rewards of reciprocity of affection or friendship.

However, gambling still is usually classified internally depending on whether one considers the risk to be voluntary or involuntary -- and reasonable depending on whether the chance of reward is sufficient to justify the amount of risk. If we think it is a high risk and does not require to be done then, and only then, do we usually call it gambling. So, games of chance are considered to be gambling but crossing the street is not.

This isn't true of everyone. Jack Nicholson, in "As Good As It Gets", portrays someone who is all-too-aware of the everyday risks of life. He continually strives to eliminate risk by isolating himself and entering into rigid routines and being hyper-careful of hygiene and exposure. The rewards of everyday life are not enough for him to take these risks. He is "fortunate" to be able to cater to these attempted avoidances of risk because he has a lucrative occupation that allows him to do this. It is only when he sees a reward that is large enough that he increases his willingness to take more risk.

Nicholson' s character is seen as abnormal because his awareness of risks is much greater than his recognition of potential rewards. The risks exist -- but so do the rewards that he cannot grasp. There is not a single, appropriate, balance even though there are certainly, within a given society, an expectation of being able to make "reasonable" judgements on such.

For someone living in a war zone, the risks of doing anything rise and the potential rewards narrow. For someone with a dependable environment and financial basis, the risks seem smaller because the downside of failure is much less even if they don't achieve the hoped-for rewards.

There are various phobias -- more specialized than those that Nicholson's character revealed -- that are still an out-of-balance reflection of the risk versus the reward. To Chicken Little, the sky may fall upon him if he goes out -- or to the agoraphobic. One person may be willing to work in high construction, balancing themselves on girders while another person may have difficulty getting onto a balcony.

Life is a gamble. Being able to weigh the risks and rewards are more computable with games of chance than they are within everyday activities. But the risks must be taken to get to the rewards. It is part of life's journey to learn to make those decisions based on known risks and benefits.

What risks do you see that outweigh the possible benefits?

82.   Out-of-Sync: When Social Evolution Lags Technical Innovation – 1/26/2017

Humans, as a whole, are very capable intellectually. They're "smart". They can figure out how to take something and improve it. They can take two or more, apparently unrelated, items and figure out something completely new that can be done by using all of them together in a way that has never been done before. This is called "technological advancement" or "technical innovation".

This is good. It has allowed us to move from caves to weather-secure buildings with running water and toilets and to be able to go from point A to point B in hours rather than weeks or months. It has allowed us to be able to feed 100 people, for a year, from an acre of land rather than 20. In most areas of the planet, it has helped us to be able to spend some time each day doing something other than trying to survive.

What humans, as a whole, are NOT very good at -- is using technology consistently for constructive purposes. We invent ways to initiate, and control, fire and some individuals start burning down forests, or houses, or even tying alternative health specialists to stakes and burning them. We figure out ways to communicate at the speed of light across the planet and use it to spread false information more quickly. We work on methods to cure disease and help those who have incurable problems to live more easily and efficiently -- then we use those same methods to create diseases that kill millions. These are examples of "social evolution" (or lack thereof 😩).

This is an example of being "out-of-sync" (or out of synchronization). Technology becomes capable of certain actions before people have matured sufficiently to consistently make use of the technology for good. Another way of putting it is that technology advances faster than people's ability to use it properly -- and the spread between these two appears to continue to widen.

In order to curtail this problem, of course, one of two solutions can be applied. The first would be to slow down technical innovation. This is unpopular and unenforceable. As stated at the beginning -- humans, as a whole, are very capable intellectually. If one government, or group of people, agreed to slow down technological development in one area then another group of people will be completely willing, and able, to proceed on the development path on their own. A similar method could be done on an individual basis (since usually there is someone who is first with a technical innovation) with that person deciding to slow down, or withhold, research or results. This might work for a while but it disrupts many social policies of reward and competition.

The other thing to do would be to accelerate social development. This sounds great but it also has problems. One problem is there no way to accelerate social development universally. In other words, if you have a nuclear bomb it only takes ONE finger to "push the button". And, within current economic patterns, there is no incentive to devote the resources (time, people, energy, "money") to help people to develop socially; there is no direct, short-term, "profit" from a more mature, socially capable, human.

So, we have a problem. The problem doesn't appear to have a solution. They why talk about it?
Because **awareness** of the problem is, in itself, an approach to a solution. If most scientists and engineers and tinkerers keep this problem in mind then technology can be presented in such ways that constructive ideas are promoted before destructive ones. This doesn't stop abuse but it does reduce it and slow it down. Such awareness can also avoid the social problem of technical abandonment -- "I just

invented it -- it's not my responsibility how it is used". A true, but very lazy, excuse.

Do you have any suggestions as to how to keep technical innovation and social responsibility at the same point?

83.   Artificial Intelligence: Beyond the Turing Test – 2/11/2017

In 1950, the British mathematician Alan Turing gave an answer to the question -- how can you tell if a machine is intelligent? His (paraphrased) response was "if you cannot tell the difference between a human answering questions and a machine answering questions then it has achieved intelligence". This Turing Test is not universally accepted but it is probably the most widely used foundation of answering the question of what is Artificial Intelligence (AI).

Alan Turing's test was based on the idea of an interviewer and a responder. Someone asks a question and someone answers a question. This led to a series of experiments in computer programs that simulated (or imitated) "normal" human interviewer/questioner situations. It might be between a therapist and patient or doctor and patient or a student and professor/teacher. Naturally, there had to be a way to make it impossible to physically tell whether it was a machine or not. It also had, built into the test, the requirement of equivalent skill in understanding and speaking/responding in a human language.

In today's world, computer programs have advanced beyond simple questions and answers. We have computer programs beating humans in Chess, and Go, (and other games). The Turing Test might not be considered to apply to these situations but many people would consider this a form of AI. We have computer program/systems that make use of pattern recognition to identify potential suspects or targets of drones. So far, the final decision is still made by humans but stories/films such as The Minority Report indicate a possibility of the machines making final decisions even about what might happen.

That is the "line in the sand" for people thinking about AI. Who makes the final decisions? Is it a human (with all of her, or his, faults and experience) or a machine (who, at heart, is still the results of a programmer's abilities and recognition of exceptions)? Isaac Asimov, in his Three Laws of Robotics, had the AI programming include self-restraints as to what the program/robot could do, or could not do, without undergoing self-destruction.

Speed and safety. The primary reason for computer programs is NOT that they can do things that humans cannot do; the primary reason is that they do things much, much, faster (and reproducibly). So, if you design an AI that handles the coordination and operation of a nuclear reactor, you want the program to be able to respond very quickly. Putting a human into the decision path slows everything down. Who has the final responsibility?

The same question exists within the possibility of self-driving automobiles and trucks. It is likely that AI programs can already drive as well as an average driver -- assuming that all of their sensors work properly (they can detect objects and highway lines and sounds and bouncing balls and the cars and buildings around them, ...). Certainly, in another five or ten years, AI self-driving programs will be able to control a vehicle much safer (and more rationally -- no road rage potential) than humans. But they would be making the final decision.

If a self-driving AI makes a mistake or a necessary decision that costs lives, who has the responsibility? The programmer? The company that built the vehicle? The owner of the vehicle? What happens if the self-driving car is involved in an accident with a human-driven car? Is there presumption of innocence on the part of the self-driving car?

In all these cases, the program and machine are taking the place of the human. If you keep them "behind the curtain" there may be no way to identify whether they are human or machine. They PASS the Turing Test. But, when the curtain is removed, what is the final verdict? Who/what has the responsibility? Who/what makes the final decision?

84.   Is Being Average "Normal"? – 2/19/2017

Many people are fascinated by surveys and tables of statistics. They can be fun -- and even useful, when interpreted by people who know what the data does, and does not, indicate. They can also be misused by presenting data in such a form that people will jump to conclusions that are not really supported by the data. This is done directly by many politicians and for political purposes and causes.

I have mentioned before the quote popularized by Mark Twain -- "There are three kinds of lies: lies, damned lies, and statistics." I like statistics because I am fascinated by numbers. Their relationships and the ability to use various formula upon the data is a lot of fun. But I don't have sufficient training to use (or misuse) them properly.

But this particular blog is not about such weighty matters. It is more about how people can take "everyday" data and make more of it than what they deserve. In particular, it is about averages and the definition of normal. An average is a fact when you are dealing with numbers. The average of two, five, and eleven is six. The average height of a human male in the United States is approximately 69.7 inches (177 cm) -- but there are not that many men in the United States who are exactly 69.7 inches tall. In other words, few men are of "average" height.

Averages only make sense for characteristics when you examine them within a distribution. A normal distribution curve looks like:

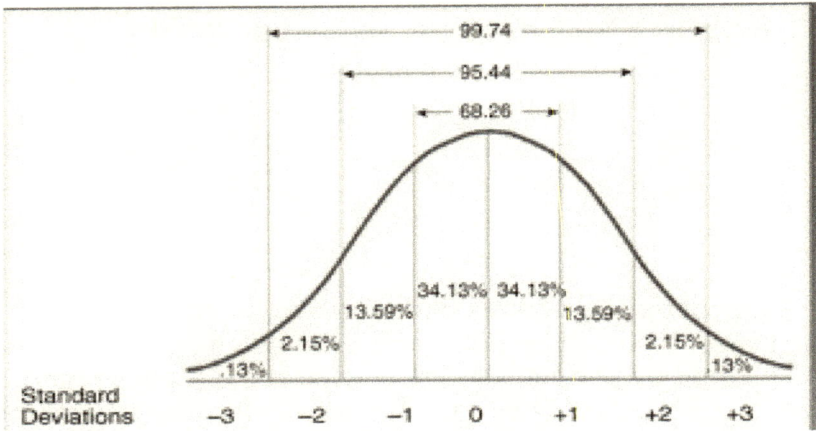

Standard
Deviations   −3   −2   −1   0   +1   +2   +3

Normal means, basically, that the curve has a uniform change throughout the graph -- there are just as many above the average as there are below the average and the percentages change consistently on both sides of the "bell curve". This ideal curve doesn't occur that often in nature. One problem that happens is that there are often assumptions that a distribution of data will meet a normal distribution without doing a sufficiently wide sampling of data to justify it. For example, the data for men and women's heights in the U.S. can look like:

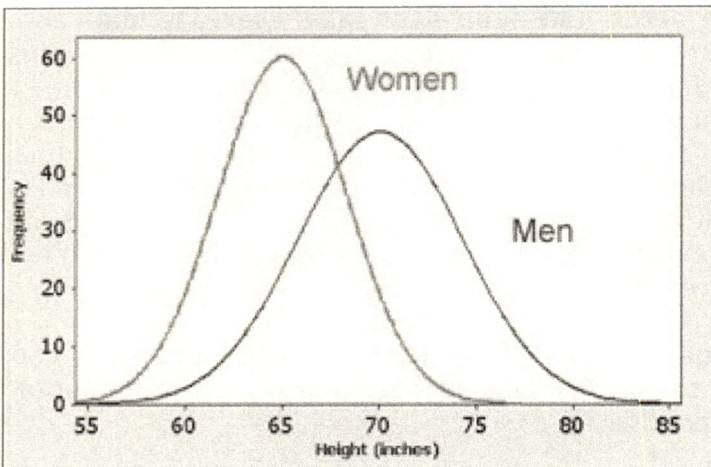

Note that the peak for men ends up around 70 inches, while the peak for women is at about 64.5 inches (164 cm). The peak is higher than for men and the curve is narrower. This indicates that women's heights don't vary as much as men and concentrate at the average. However, in both cases the curve is very "smooth" and symmetrical. For height, it seems that the data really does seem to support this -- with a longer "tail" at the tall end indicating a slightly greater number of very tall people versus the number of very short people.

Any characteristic that can be measured can have an average. Skin color is based on the amount of pigmentation from different types of melanin in the skin. There could be an average value for this. The average weight for a U.S. male (in 2015) was 195.5 pounds (88.5 kg) -- an increase of 30 pounds (13.5 kg) since 1960. The number of hair follicles per square inch can be measured -- thus, there is an average value for "hairiness". Ear size can be measured from top to bottom (or front to back or distance out from the skull).

How many people have an average height, weight, skin color, hairiness, and ear size? Probably only a half dozen or so within the United States. Are these six people the "normal" ones? No, not even from a statistical sense because there are going to be many other measurements that are potentially able to be done -- eye color, IQ, foot size, hand span, distance from bottom of nose to top of lip, and so forth. Everything that can be measured can have an average -- but none of them are "normal" because humans are not just one characteristic. They are combinations of many, many characteristics and there may be some that are genetically linked (if you have one then you also have the other) but most appear to be totally independent. Skin color is totally independent of IQ. Eye color appears to have little to do with height. Weight and height do have some correspondence but it is possible to have a tall thin person as well as a short heavy one.

Since being average in every possible measurable area is highly unlikely, it is certainly not normal to be average.

85. Synchronization: One is a Wiggle, Multiples are a Movement – 3/4/2017

We can sit back and watch someone dance. We observe the fluidity of their motion, the deliberateness of the position, the rhythm of the movements. Perhaps it is even in accordance with the components of music -- the tempo, the syncopation, the tones. The body becomes an extension of the mind and spirit -- an instrument for expression. The experience is different for the participant than for the observer but still capable of extending beyond words, or sounds, or sights.

But, what happens when we observe someone (or do it ourselves) who does something that is considered "clumsy"? Something is dropped. The body sways to one side in an erratic way. A jump is made but the landing does not meet expectations of how it should be done. There is a fall, a jostle, a slip. Instead of smooth transitions, each movement is jerky and without observable rhythm.

We have probably all noticed this -- in ourselves and in others. Perhaps we have also made judgements about ourselves. "I just can't dance." "I have no rhythm." "I cannot let myself be seen in public."

But what happens when movements match within groups of people -- or even couples? One person raises their arm up at a 45 degree angle. What is going on? Two people raise their arms up at a 45 degree angle at the same time. Does it feel different? How about when a dozen people do it at the same time? It can no longer be considered an accident or a coincidence. There is synchronization -- individuals are becoming part of a group and using that coordination to communicate something.

What are they communicating? Ah, that is a question indeed. About the only answer I can give is "42". Bees join into groups of movement to indicate food supplies and directions and weather and other things vital to the colony. Aliens from other worlds might join to indicate similar things -- so could humans, if desired.

What about our examples of "clumsy"? What if two people sway to one side at the same time? What if an entire group of people jump up and land in an unexpected way? What if the jerks happen in a series of movement within a group of people -- no longer at the very same time with each person but in a series that is apparent even if not predictable?

What we can see from this is that movements do not carry inherent worth -- or they all carry inherent worth. While an individual may tune their body to express in ways that are socially recognized as approved movements, the individual movements that are not so approved can still be appreciated when synchronized within a group.

So, am I clumsy or have I just not found the group to match my wiggles?

86.   The Magic Penny Effect: Why Greed Causes Economic Disruption – 4/23/2017

Malvina Reynolds, an American folk/blues singer-songwriter, wrote (among other things) a song called "The Magic Penny". You can see a full set of the lyrics at The Magic Penny song but the first verse and chorus go:

Love is something if you give it away,
Give it away, give it away.
Love is something if you give it away,
You end up having more.

It's just like a magic penny,
Hold it tight and you won't have any.
Lend it, spend it, and you'll have so many
They'll roll all over the floor.

As talked about in my earlier blogs on money and economics, money is a symbol of resources. You cannot directly eat money, or grow money, or save money -- money is only a symbol. The symbol can take many forms -- solid ones such as gold or other "precious" or rare metals, electronic ones such as bitcoin, paper ones such as pound notes or Euro notes, or solid symbols of other wealth such as physical coins.

Initial use of money arose out of the difficulty of having precisely what the other person wanted as barter -- or the difficulty of transporting it, keeping it alive, and transferring it. Even in a regular barter economy, it is difficult to have that cord of firewood in your pocket if you want somebody's fish. Much easier to have some mechanism of recording that the woodcutter owes you a cord of wood at some point in the future. Even easier if there is some common unit such that one cord of wood is equal to five Tunkels and one fish is equal to one Tunkel. Therefore, one cord of wood is equal to five fish.

The magic penny effect is most directly related to the practice of **hoarding**. When you hoard something -- whether it is money, or food substances, or newspapers, or whatever -- you take it out of "circulation". It is unavailable to be used. In the case of food, it will eventually go bad and be unusable by anyone. In the case of newspapers, they can rot and the information will become outdated. in the case of money, those symbols of resources disappear from the economy. While they are not actively used, they are the equivalent of not existing.

As the song goes, you can spend it or lend it and it will be an active resource. Present, and usable, to convert into food, or housing, or video games, or whatever. Within a capitalist society, it can be loaned to those who do not possess adequate symbols of resources at present and a tax (interest) can be charged against what they can contribute in the future. But, if you "hold it tight" it serves no purpose and might as well not exist -- and has the potential of disappearing (stolen, lost in an earthquake, paper equivalents burned, etc.)

This is related to, but not the same as, the build-up of "phantom resources" or "accumulated capital". Controlled by a central center -- and able to be spent or lent -- but not freely available to those who can most use it (or, potentially, most deserve it. We'll plunge into that topic in a soon-to-come blog. But, for now, keep in mind the "magic penny" effect and see that money is actively used.

87.   Spillover: When Trickle-down Meets the Dam of Unregulated Income – 4/29/2017

About three years ago, I wrote a blog about the "trickle-down" economic concept (Trickle Down Pyramid blog). In that blog, I talked about the difference between the advertised behavior of how the concept is supposed to behave and how it behaves in reality.

However, at that time, I did NOT talk very much about how this split between concept and reality occurs. I will try to go into some of the aspects of how lack of regulation and proper progressive redistribution (whether via taxes or other method) makes this inevitable.

Trickle-down seems to make sense (see the first distribution table from my earlier blog) as a concept. The idea is that you give resources (labor/money/etc.) to a set of people who can, and will, make use of it to increase productivity and the general welfare. In this process, they distribute the resources to a second "tier" of people who do the same thing by passing along to a third tier, and so forth. Thus, the resources are funnelled through the top of the pyramid and "trickle-down" to the rest of the population.

This theory is not actually unique to capitalism. The same theory holds for many other economic philosophies. The differences lie in how the group is chosen to start the distribution and the rules of distribution. In the case of capitalism, the idea is that the people who have accumulated the most capital are the best at making use of the capital. In unregulated capitalism, these people are allowed to disperse the resources/labor/money as they see fit -- with the underlying expectation that, having previously used capital in such a manner to increase it, they are best qualified to continue to do so. Remember that money is only a symbol of resources -- so by increasing money supply they are, in theory, increasing the amount of resources (food/labor/clothing/health/etc.) I will use the world capital in the rest of this blog as a shorthand term for money/food/resources/clothing/health/etc.

Looking at the above definition and reasoning, a number of potential problems are apparent. It is also true that, IF the person or group is competent and trying to distribute, and increase, resources that deviation from unregulated capitalism is not needed. Alas, that is very rarely the case (similar to the idea that an intelligent, benevolent, dictator can be the most effective and efficient -- but such people are very scarce and almost never succeed in continuing the system beyond the life of the original).

The first potentially erroneous assumption is that the people who have accumulated the most capital are the best at making use of the capital. This is true only in the cases where they have demonstrated that they have these skills by starting with very little capital (perhaps only their own labor and ideas) and building it up by the proper use thereof.

Note that, even in these cases, they are not using ONLY their own capital -- they are making use of lots of public capital (roads, energy supplies, education systems, health systems, safety systems, etc.) However, these people do demonstrate that they can make use of their own capital, in combination with public capital, to increase and accumulate. They do so with an inherited debt to the public for the contributions of the public capital.

Can those who inherit capital have the skills? Yes -- but it is impossible to demonstrate or be certain of it. At the best, it is a different set of skills to those that are used from building up from initial levels. At the worst, it is a completely parasitic relationship -- where they are taking from the capital base and actually decreasing the distribution and use. It is "unearned" income and, arguably, undeserved. Possessing inherited capital does not indicate any ability to properly use it for the sake of the general population and should be limited as much as is possible within the system.

The second potentially erroneous assumption is that they will make use of the capital (resources/money/labor/etc.) in a manner such that it will create MORE capital. This is the second aspect of "trickle-down". Capital has to move and be used. As in the previous blog, it cannot be held onto; it cannot be kept without circulation. Furthermore, it must be distributed to others who will ALSO be using it to create more capital. This is what creates the additional "tiers" which allow trickling to take effect.

In summary, trickle-down runs into three specific problems. These problems are of inappropriate allocation and inheritance, retention, and lack of leveraged distribution.

We will go into further detail as to how capital is properly distributed in another blog. However, if a group or individual is displaying a "lavish lifestyle" then it is NOT being properly distributed. As described in my blog of three years ago, the retention and personal use of excess income distorts and damages the economy.

88.   Income: The Class Structure of Democracy – 5/5/2017

There are a number of topics that tend to make people uncomfortable to discuss. Among these is the reality that there is a class structure within democracies. Some of the original voluntary immigrants to the U.S. were seeking the mobility of opportunity without requiring to be born within the right hereditary class of their society. Democracy is one of the "great equalizers" with one person -- one vote. Unfortunately, it isn't quite that straight-forward -- I'll tackle that within a future blog. But, whether we want to acknowledge it or not, income forms distinct classes within democratic society.

It is generally recognized that there are three broad categories of income. These are Upper income, Middle income, and Lower income. Sometimes the word "class" enters into descriptions of "upper class" and "middle class" -- it is rarely used with lower income. I would propose that these three categories can be summarized as Survival, Stability, and Search for Purpose. Each of the three primary categories can be broken into three sections: bottom, central, and top. These are categorized by emotions of fear, complacency, and hope; there is fear of dropping from one category to the next lower, a complacency of "deserving" to be in a category, and hope of rising to the next category.

People have the greatest knowledge of the income sections that they have directly experienced. They can also "glimpse" into sections that are adjacent to those within which they have lived. It is indeed very difficult for people to "walk in another person's shoes" without having **lived** that life and, thus, there is great difficulty for people to truly understand the lives of those two sections away from what they have experienced (which is an excellent thing to recognize when voting).

For example, someone who has lived their life within the central upper income section doesn't know, and cannot directly understand, the lives of those in the middle or lower income categories. In a different fashion, someone who has lived at a bottom lower income level can only fantasize about being in the middle or upper income categories. Very occasionally (and less and less frequently), those fantasies become dreams and those dreams become reality. But they always remember (sometimes with fear and sometimes with compassion) where "they came from".

There is also a comfort level that is felt within the economic section in which one grows up. Consider the "Beverly Hillbillies" U.S. TV show where they become rich but still want to live the way they did. This is a story that is often told about many people who have become rich (performing stars, sports celebrities, etc.). Of course, it goes the other way also -- where people who have lost their wealth can never be content or comfortable at a lower income level. This has an inadvertent side-effect on those who win at lotteries -- they are neither comfortable with, nor are able to properly use or invest, their additional moneys and often end up losing it all.

During the Great Depression, there were photos taken, and articles written, about wealthy people who committed suicide after having "lost all they had". The reality was that they had usually lost 70 to 90% of what they had -- plummeting them down three or four income sections. The amount they had left would have been considered adequate, or even attractive, to someone in the middle income category but, to them, it was a fate that could not be contemplated.

My mother was born in 1930 and my father in 1933. They were fortunate that they were born into farming families -- away from the center of the effects of the Dust Bowl. They had food and didn't lose their housing. However, my father was three years old when he lost his own father due to pneumonia and so his early life was also a matter of surviving through the Great Depression. This experience provided a foundation to their life and, indirectly, provided a foundation to mine.

My father completed ninth grade in high school when he left home. My mother finished tenth grade. Both were hard workers and did well in school but the needs of the family meant that continuing school was not the most immediate priority. As I was growing up, we vacillated between "top lower income" and "bottom middle income" income classes. As I did complete high school and college, I would probably be considered to now be in the "top middle income" region. So, I have either lived, or been able to "glimpse", six income sections. It is a core part of my self-image and relation to my economic life. Each week I am grateful for having a job, somewhat fearful of potential loss (to which I react by saving), and continuing to recognize the daily situation of those in the income sections below my current one.

My children have only known central middle income and top middle income -- about which I am occasionally sad because they have no direct knowledge, or understanding, of people who have lived within the lower income sections. They do have plans to help join a summer work program to help within a bottom lower income community this year -- perhaps they will gain some insight at that time. The Peace Corps, and similar programs, require people to live at an income level close to that of the communities within which they serve -- perhaps that should be a requirement for all people who wish to run for public office.

Each of these sections can be categorized by particular behaviors and challenges. However, this particular blog is becoming a bit on the long side. If you are interested in more detail, please tell me so.

89.   Language: Translations of Thought – 5/14/2017

I have always loved language. So far, in decreasing order of proficiency, I have studied English, French, Russian, German, Spanish, Farsi, Mandarin, and Italian (currently working within Mandarin). But don't expect me to be able to immediately respond in anything but English (and, perhaps, French) as I have never been in the situation of actively using languages other than English on a daily basis so, over my 60 years of language learning, there are some parts that will need an archaeologist to unearth. That doesn't stop my love of languages or my desire to understand.

It is sometimes still a puzzle for me to move between what language indicates and what it means but I still love it. I cannot stand euphemisms, however, because I feel like they steal from the language. They are, in my opinion, the "anti-language" whose purpose is to create confusion and make communication more difficult. I will tackle them in another blog, as they deserve their own piece.

A lot of people like to categorize -- and I confess to being part of that group (in case you have never been able to tell from my blogs). Languages can fall into two general groups -- "natural" and "created" languages. Natural languages develop from the first grunt up into a sequence of complex sounds and structures that are unique in relation to the world. Created languages are devised by communicating individuals for specific purposes, with specific goals and (usually but not always) specific audiences.

Natural languages do not appear only within the human world. I have recently read an article about the creation of a bottlenose dolphin dictionary. Jane Goodall, within her works with chimpanzees, has used modified sign language (a composite of natural and created language forms) to communicate with the groups of chimpanzees -- but the chimpanzees do start with their own language directly suited to the needs of their environment. To the best of my knowledge, created language remains within the world of tool-manipulating sapients but not necessarily just humans.

Created languages include those to communicate with machines and those used for specific purposes. For example, Semaphores are coded forms used with sight and hearing to allow simple translations of existing meaning into a form that can be transported over a distance. Within the world of computers, there is machine language which is composed of "instruction sets" which are directly acted upon by the components of the processor. These are built upon by assembly language which is a more readable -- but directly substitutable -- form of machine language. And upon this lies the "high level" computer languages which translate into reproducible larger sets and structures of assembly/machine language.

Within high level computer languages, the designers have the opportunity to use the language as "shorthand" for things they want to be able to accomplish with fewer commands. SNOBOL is oriented towards manipulating "strings" -- such as sentences or word patterns. FORTRAN was developed primarily for executing mathematical FORmula TRANslation. 'C' is sometimes called the most low level high level language there is because it allows the greatest direct equivalencies to assembly language (even allowing direct insertion of assembly language within a program) without directly using the structures of the processor.

Other created languages have been formed for social reasons. Esperanto was designed to be a universal human language. There is now an active Klingon language (prior to discovery of physical Klingons) based on the ingenuity of enthusiastic fandom. Pidgin, or "trade" languages, have been created as a merging of the dominant language of the traders and the local language of the people with whom they want to trade.

Natural languages arise first out of the physical environment. If you live in the Arctic region, then snow and ice are important aspects and need, and deserve, different

272

words variant on the type, use, method of creation, and other characteristics of the frozen water. On the other hand, if you live in the middle of a desert, words for frozen water may not appear very soon but words for types of heat, sand, and moving terrain would be more useful. Vocabulary from life in the high mountains would be different from that of the lowland tropical forests. Your language must first reflect the world around you.

Beyond the physical environment comes the social, cultural, political, and religious world. In the Russian world and environs, where a central authority has been in control for more than a thousand years, the language reflects a directive authority towards the larger population. People do not do things -- things happen to people. German acts as a "push down" language where many concepts, items, identifiers, and actions may be present and kept as part of the overall context until the sentence has been completed -- at which time all of the parts of the sentence interact. In the early forms of some languages such as Japanese, there are two languages co-existent based on what can be done, and thought, within different social classes.

An acquaintance pointed out how well suited the ideographs of Asian language are for empires and acquisitions -- the written language acting as a common foundation under many different verbal variants. In addition, the written Chinese language puts together thought concepts into more complex ideas that then take different forms within context. This is very different from written language which reflects the phonetic (sounds) aspect of a spoken language. In the one case, if you can read you can speak. In the other case, if you can read and write you can move concepts from one group to the other without any commonality of the spoken language.

The recent movie "Arrival", is based on the short story, by Ted Chiang, "The Story of Your Life". In this story/film the language is independent of time and, in turn, allows the mind that understands it to be free of the constraints of linear time. Although this may appear to be beyond likelihood of reality, it is true that language gives structure, and constraints, to thought while thoughts are difficult to express without a means to express ideas held in common between two, or more, people.

Language is a reflection of the history, culture, and story of a people and there is loss when the languages are lost. Translations can be approximate but not exact. But, like with the ideograph, it is possible to create a meta-language that is a concept-based set of structures to allow better translation from one language to another -- which is one approach that current Artificial Intelligence (AI) programs take. No matter what is done, languages open many doors.

90.   The Puritan Effect: Why are U.S. Jails so Full? – 5/20/2017

In the United States, our incarceration rate is number one among the larger countries (The Seychelles have a slightly higher rate but with many fewer people). So, why does the U.S. have so many people in prison? Are U.S citizens just natural crooks?

And should we care? Well, the greater the number of people there are in jail, the fewer there are paying taxes and participating in constructive contributions to society. In addition, a person in jail costs the community/state/federal government from $70,000/year to $200,000/year. Personally, I would much prefer to use that kind of money on improving education, infrastructure, child care, and other public benefits. Remember that, even though there is no such thing as a "free lunch", we SHOULD be able to determine how our tax money is used.

The reality is that the incarceration rate varies for all countries depending on various factors -- some of which fluctuate because of political and economic events. In the U.S., we had a very low rate of imprisonment in the 1930s and 1940s -- possibly because the U.S. economy could not support that many people in the prisons so it was primarily the violent offenders that remained there. In the 1960s, the U.S. had a "world average" of imprisonment and it has continued to rise until, today, it is about the highest in the world with around 700 people in prison for every 100,000 citizens. In contrast, Cuba has about 500, Russia has about 450, Costa Rica has about 350, and the United Kingdom has about 150.

People are basically the same all around the world. Some are jerks. Some are criminals. Some are saints. Some are lazy. Some are hard-working. Some are poor. Some are rich. Some are smart. Some are not smart. And so forth. External aspects don't make much difference -- skin color, gender, religion, nationality, et cetera. There are some cultural influences on how important studying may be or how much physical activity is praised and rewarded but, at heart, everyone is in the same human pool.

So, why does the U.S. presently have such a high prison rate? A large part of it has been the slide of income inequality as (true in every country) there are a higher percentage of poor people in prisons than there are of rich people (once again, NOT because poor people are more likely to be criminals). Keep in mind that a "crime" means that a law has been determined to have been broken. Thus, something may be a "crime" in one country, or by certain people, or at a certain time -- and NOT a "crime" in a different country, or when done by different people, or at a different time.

But, specifically, there are a number of factors (not firmly listed in order -- but perhaps in order of most easily changed):

- For-profit Prisons. This makes absolutely no sense within a capitalistic society. Capitalism operates (loosely) on Supply and Demand as well as profit margins. In order to increase profits per inmate, a prison will work to reduce expenditures which is likely to increase the chance of the criminal coming back. Also, it is to their interest to INCREASE the number of criminals and they do this by heavily lobbying for increased sentences for a greater number of newly created "crimes". A profitable prison means a greater and greater number of criminals every year.

- The Pilgrim Effect and Vice Crimes. The Pilgrims were a small group of settlers within the United States -- very small in number in proportion to all of the other immigrants (all in the U.S. are immigrants -- even First Nation). But their effect on the national psyche is enormous. Or, at least, that is the way I refer to it.

The Pilgrims felt that they should control how every individual behaved and thought on a daily and minute-to-minute basis. Politicians in the U.S. make use of this desire to control others by creating laws (and, when laws are created, so are crimes and criminals) that are aimed at the thoughts and behaviors of citizens THAT DO NOT AFFECT OTHERS. These are also called "vice" laws. There are lots of them -- and a huge difference between the U.S. and other countries in longer governed countries. Laws about drugs (including alcohol), gambling, sexual behavior, and so forth.

One problem with these laws is they do NOT work. The "vices" continue to happen. Secondly, they do succeed in increasing the profits of those who are involved with these activities -- causing additional corruption and organized, highly profitable, criminal businesses both within and without the U.S. The U.S. attempted "Prohibition" of alcohol with the 18th Amendment. After giving the Mafia, and other organized crime, a huge boost in the U.S with attendant increases in murder and other violent crimes, the 21st Amendment reversed it but much of the damage to the country remained.

A side-effect of "vice" laws and increasing their profit margins is that people who feel the need or desire for these substances/behaviors have a greater need for money to participate. This causes a multiplying effect when robberies, assaults, and other property crimes are done to be able to afford the law-enhanced prices.

- Income Inequality. There are a higher percentage of poor people in prison than there are of rich people. Some of this is similar to the predicament of Jean Valjean (Hugh Jackman in the recent movie version) in Les Miserables -- turning to crime to feed their families when they are desperate. By definition, the rich are almost never desperate except in non-economic situations.

Secondly, most of the laws are written BY the rich

and FOR the rich. In the federal government, most members are millionaires and it is unlikely to be coincidental that the longest sentences and greatest number of "crimes" are created for actions more likely to be done by the poor and the lightest sentences and fewest "crimes" are created for actions more likely to be done by the rich. The income discrepancy lessens as one moves down to state and local levels but still exists and makes a huge difference.

- Recidivism and Employment. When "Antman" gets out of prison for a non-violent, capital property crime he gets (with his high-tech degree) a service job from which he is fired once they find out he was convicted of a "felony". While this may be a bit out-of-place (high-tech people are less likely to be discriminated against) the general attitude, and situation, is NOT. It is as if a "felon" is given a lifetime sentence as they are forever actively prevented from being a positive economic or societal force. They may literally have "no choice" but to get sent back to being taken care of in jail.

  In addition to permanent economic retribution, ex-prisoners (especially from the for-profit prisons) may not have improved their ability to get a job (assuming anyone will let them have one) and be sent back to the same economic and societal environment that left them feeling that "crime" was their best option.

- Racism. J. Edgar Hoover, head of the FBI from 1935 to 1972, was a fervent racist and implemented dual-prong attacks to help keep blacks, in particular, from achieving greater opportunities. He instigated additional drug sales into poorer neighborhoods while, at the same time, blackmailing and pressuring Congress into passing laws making use of those same drugs to be a crime. What didn't use to be a crime -- now was. And what didn't use to be a problem in those neighborhoods -- now was.

278

Prior to the various Civil Rights laws and, once again, since the Supreme Court decided (on June 25, 2013) to remove the ability to enforce many of those Civil Rights laws, many laws were put into place where they applied ONLY to people of certain definitions -- usually of certain skin colors or of more recent non-Anglo heritage.

- Unequal Enforcement. Related to income inequality in that there are different laws defined to be more likely oriented towards the rich as to the poor. However, even within the same category of laws, certain groups are favored over other groups. These groups include "whites", the rich, males, and other majority (or formerly majority)-oriented groups.

A poor person may get 10 years of prison for a $10,000 theft. Meanwhile, a rich person may get 6 months probation for stealing $10 million from a group of people. The punishments are actually reverse proportionate. Discrepancies get even greater when corporate crime is committed, with no one actually getting punished at all (and only a mild discomfort when any fines are passed along to the employees and stockholders). This is a broken Justice System.

Note that, in the United States, there is the notion of a "jury of our peers". This should mean that poor people should be judged by poor people and rich people judged by rich people -- but doesn't always quite mean that. However, poor people are more likely to be harsh with other poor people. I believe it is an attitude of "I had to struggle so hard to survive without having to commit crimes, how dare you take an easier route".

- Alternate Punishments. Of course, one method of reducing the number of people in prison is to have alternate punishments. It is possible to have periods of public work labor. It is possible to have a separate "withholding" of parts of a paycheck(s) to pay back damages or theft losses.

The idea of alternate punishments is to address the damage(s) caused by the "crime" while retaining the ability for people to continue to contribute to society. This is normally only possible for non-violent crime. In some countries, prison is avoided by increasing the frequency of the death penalty -- which I don't recommend because it is so difficult (impossible) to correct mistakes (which do happen on a regular basis).

In summation, the U.S. could quickly reduce the number of people in prison. Eliminate for-profit prisons. Eliminate laws that are about "crimes" that only affect the individual -- the "vice" crimes. AND make it retroactive such that all who have been convicted under such laws are released and their records eliminated. Make it a requirement to have a reason-produced court order to obtain criminal records of job applicants.

Other aspects of imprisonment imply a direct change to culture and society that may end up taking many years -- but still are worthy of effort for change.

91.   Elections: Why Does Money Matter? - 5/29/2017

In the United States, we have just finished a major election cycle but elections never actually end. There are "special elections" which are held to fill positions that are vacated -- whether from causes of illness or death or that of being shifted to another position, either elected or appointed. There are local elections and state elections and special issue elections. And, there is the continuous preparation for the *next* election cycle.

During any of these, unless you are remarkably successful at making yourself invisible, you will be approached by one campaign or another (sometimes both sides, or multiple candidates) to make a donation. And, if you do make a donation, you may feel assured that you will be approached again ... and again ... and again. In the age of electronic distribution of information, this may well often take the form of social media postings and/or electronic mail. Always a crisis. If the polls are "up" then this is a time to solidify the lead. If the polls are "down" then it is desperate for more money to combat the positions of the other candidates or supporters or opposers of issues.

Money, money, money. Why in the world does it matter? It certainly should NOT matter -- an election should not go to the highest bidder within a democracy. Yet, except in a relatively small portion of elections, the person/issue that spends the most money is likely to win. Why should this be so?

To delve into that answer, we first need to remember that what we call a "democracy" is almost always actually a "representative democracy". In other words, we do not vote on bills, or articles of government (we sometimes do get to vote on issues or referendum) -- we vote for people whom we trust to **represent** us. These people are supposed to have similar views and opinions to the majority of the voters who elected them.

This is not easy. Even in the 1800 United States Census, there were counted as being 5,308,483 people (including slaves) in the United States. With 106 Representatives, that means that each Representative, on average, was to present the points of view of 50,080 individual people. (Of course, not all of those 5,308,483 people could vote -- but they were still supposed to be represented.) There were sixteen states, so there were 32 Senators -- giving a ratio of an average of 165,890 people represented by each Senator.

So, we see that, even in 1800, there were too many people per representative for the people to know them well. (This is also true about the Electors who were appointed or elected to represent the people at the Electoral College -- another potential blog.) So, how to know? (Note also that, at present, each Representative represents approximately 740,000 people and each Senator an average of 3.2 million people -- so the ratio has not improved over time.)

There are two basic needs for someone to be considered appropriate to represent people. The first is clear, although not necessarily obvious, and that is that her, or his, name must be known and recognized for which to vote. The second is that there must be a perceived agreement of values and judgement such that this person would be whom we would choose.

On the first area, the requirements of the present are similar to that of 1800. Their names are known because they are talked about. George Washington, John Adams, and Thomas Jefferson were probably "household names" due to participation in the various needs of the drive for Independence. Later Presidents (and the opposing candidates) were not necessarily known from the American Revolution but they were known for various events (wars, other political matters, etc.) that appeared in newspapers, or the neighbors talking about someone who knew someone who knew them. The media spread their names and the neighbors spread the opinions and discussions.

Within the present day, the media has become much more important -- including what is referred to as the social media. Unfortunately (in my opinion), full and frank discussion with neighbors is much less likely. These are methods of spreading the name.

They are also methods for providing information about the candidates. Not necessarily truthful information but information. There are two general methods of getting information -- "push" and "pull". When people are being "push"ed information, they are receiving information from others who are interested in you receiving it. On the other hand, if you are "pull"ing information, you are in control of what information you want to receive and also you are in control of the sources. While it is completely possible that information that is "pull"ed from an information source is not accurate or truthful, there is control by the recipient both in the type of information and the various sources of information (and there **should** be multiple sources).

OK. Now, what about money? Isn't that the core of this blog? Information is spread by the media. The media chooses to spread information because they think that people are inherently interested or because they are paid to do so. The first is called "free publicity" and every candidate attempts to get it (some are vastly more successful than others). Some, however, must be paid for -- bumper stickers, yard signs, lapel pins, and so forth. In addition, they may have either live (people -- sometimes paid and sometimes volunteer) or automated ('bot) banks of telephone lines to call people to pass the information they want to send. There are also events to which the candidates may travel to pass their words along "directly" to people. Note these are all "push" types of data transfer. And "push" rules the day in the case of elections. Sometimes, the amount of information becomes a bombardment with so many (often contradictory) items of information coming that the receiver just stops listening.

What about "pull" information? Well, most candidates have campaign offices and campaign websites (presently -- none existed in 1800) and this lists the official positions of the candidates. However, like the "push" information, this is information that the candidates want you to have (truthful, not truthful, or in-between and misleading). In order to get information that the candidates do NOT want you to have, it is necessary to go on other websites and check other sources of information. When a claim is made, check that claim from **multiple** other sites. Check the government records for recorded votes on issues. Don't be particularly surprised if you find that the candidate's position on various issues differs from the way they vote.

Is it surprising that most of the information spread by the candidate is of the "push" variety (only a limited amount of "pull" information publicly accessible to interested people)? Unfortunately, it should not be -- candidates want you to absorb (and believe) information that they want you to have. The only way that you can increase your chances of getting accurate information is to "pull" the information from many different sources. This takes time. This takes energy. And, most importantly, this takes an open mind -- which is often very difficult to maintain after all of the bombardments of the "push" information.

So, in summary, most of the money goes for "push" information. The exact expenses include administrative, equipment, events, media, payroll, strategy, research, technology, travel, and others. The emphasis on "push" information means that they want you to base your decisions on the information that they give to you (including the information that their opposing candidates produce on them). It is not good to rely solely on "push" information but it is the reason that money makes such a huge difference -- the majority of people who vote rely mostly on this "push" information and money allows this to reach as many people as possible and as often as possible.

If most people did their own research, then money needs would be limited to name recognition, "pull" sites for information, and a small amount of "push" money to allow people to see/hear/touch the candidate to make that closer feel. But "push" information rules -- and so does money.

92.   Comparison Shopping: When No Choice Seems to be a Good One – 6/24/2017

I am a consumer. I admit it. I also try to make conscious choices -- picking products from companies that are less harsh to the earth, that work constructively with their employees rather than against them, and so forth. These are characteristics of the companies that are important to me that lead me to consider products in the first place.

But, beyond that, I want to pick the product that seems to be the "best" to me. In order to do that, I need to understand what qualities I want, or feel that I need. Once I understand that, I check reviews and product comparisons to see what products best match to my desires. It is a logical process but it can drive people around me totally nuts if this is not the way they approach purchasing items. (The way they approach evaluations also drives me nuts -- it is an equal opportunity situation 😄)

In what is referred to as a "free market" situation, it is possible to make decisions solely upon the perceived qualities of a product. Qualities can include appearance, durability, price, features, societal responsibility, and so forth. Note that the perception is what counts. If a societal section decides that orange is the most beautiful color for eyeglasses then that is it. If a tall person is looking for a car then the head and leg room of the driver's seat area will be of great importance and value. If a short person is looking then they will consider a completely different selection of cars based upon their own needs -- probably for seat adjustment and ease of access to hand and foot controls. Each person has different needs and, thus, different sets of qualities.

Some qualities should be more objective. A new tire with a lifetime of 50,000 miles should be considered to have better wear than another tire that has a lifetime of 20,000 miles. Of course, there may be other qualities that are more important than lifetime -- such as the ability to handle wet roads in Seattle (which may be of no interest in the Sahara).

But what happens when you have only two choices of tires -- one that has a lifetime of 20,000 miles and one that has a lifetime of 15,000? All other qualities equal, people choose the tires that last 20,000 and the manufacturer of the 15,000 mile tire goes back to their labs and tries to develop a tire that lasts 25,000 miles. This is called constructive competition. The end result of such competition is the development of better products and lower prices for consumers. This is the ideal situation for the consumer.

What happens when there is only ONE (1) supplier of tires for a region? They sell 20,000 mile tires for a while and then they decide to start selling the 20,000 mile tires for a higher price and selling a 15,000 mile tire for the same price as they used to get for the 20,000 mile tire. This is called a monopoly situation and is not a good one for the consumer or the quality of products. Monopolies can either be established by "natural" restriction of access to resources (all of the widgets are found only in areas controlled by a company) or by economic leverage (all of the competitors are under-priced until they go out of business or are purchased by the larger company). Laws can be created to control this situation to improve the situation for the consumer and for the improving qualities of the product -- but sometimes no laws are created and the monopoly continues to exist and degrade quality and inflate prices.

By definition, there cannot be a monopoly with more than one company in active competition. But there can be private agreements between two or more companies that allow both, or all, to expand their profits at the expense of the consumers and which produce no constructive competition. This group of two or more companies can be called a "cartel". Although no single company within the group is a monopoly, they are able to control access to resources and to the consumers so that other companies that are not part of the cartel cannot compete. Laws can be devised to restrict this but they are much more difficult to monitor and enforce -- and, once again, it is possible that no law will be created.

Finally, it is also to the advantage of each company to restrict comparisons. This can be done by restriction of the movement of information (control of media access, for example). It can also be done by creating "brand loyalty" -- such that the consumers directly identify with the product of the company rather than the qualities of the product. With sufficient brand loyalty, a company has a base of consumers for which it needs to provide neither quality nor value.

So, on to the title of the blog. What happens when no choice seems to be a good one? In a free market situation new companies will arise if adequate constructive competition does not take place among existing companies. Old companies die away if they cannot maintain satisfaction of their consumer base. This can only happen in a truly free market and a truly free market can only exist with regulation and supervision. Otherwise, monopolies, cartels, and other restrictions penalize the consumers and the quality of the products in favor of the profits of the companies.

I used tires as an example of a product. In reality, anything that people choose (or do NOT choose) is a product. This may include political candidates, or services (education, for example), or resources (water, for example), as well as manufactured products. Within any product marketing situation the free market allows consumers the best options. Giving the illusion of a free market while actually controlling the choices of the consumer favors the producers. It is a continual struggle as the needs of each are weighed against each other.

93.   I Am Not Getting Older: I Am Acquiring a New Layer – 6/30/2017

I have the privilege of turning 60 years of age this year. It is indeed a privilege but I cannot truly claim that I feel like I have been growing older -- it's just that more and more of the people around me are growing younger each year. While there are indeed parts of this body that are not running as well as they used to, there are a (too) few parts that run better than ever. I am, indeed, the same person that I used to be but, like a tree with its tree rings, I keep accumulating layers.

Perhaps an analogy to a Matryoshka nesting doll might be more appropriate. Someone from the outside notices only the outside layer but I have access to all of the layers down to the first one that became self-aware. (That seems to vary from person to person -- some even claiming to recall their birth.)

That doesn't mean that all of the layers are intact or equally accessible. As is true for most people, there are periods of my life that didn't go so well -- or had outside events happening that prevented (or accelerated) growth during that time. In this situation, perhaps tree rings are a better analogy; growth rings get wider or narrower depending on existing conditions for the year. Perhaps, like the view of light in physics, the layers can be seen differently depending on what is being observed.

This view is not symmetrical -- people on the "outside" first see the outside layer of the nesting doll. And, frankly, people are not very good at observing those layers. Due to wonderful inheritances from my parents and grandparents, I got my first white hair at age 11 and was largely bald by age 30. A teenager gave me my first senior discount at age 38. (I didn't ask for it but who was I to embarrass them by refusing it?) It works in my favor now, however, as people start saying how young I look for my age.

I say that people first see the outside layer -- but once people start getting to know others those layers start peeling away. I have an advantage, of course, because I already know what is underneath my own layers. Younger people may have more difficulty finding my layer that is parallel to theirs. One has to be a bit careful in "behaving one's age" if, for that day, you actually feel that 21-year-old within you trying to peek back out.

One of the exercises of advancing age is to keep de-nesting the layers of the doll. If a situation calls for the attitudes of a 21-year-old then do so (tempered by the (hoped-for) wisdom and experience of your later layers). It is important to not let the layers get stuck together because there is so much that was learned and times when the attitudes and approaches of each age are the most appropriate.

94. Automation, Employment, and UBI: How Do They Ineract? – 7/8/2017

In the early 1800s, groups of English craftspeople banded together (in a loosely organized fashion) into groups called Luddites. They were primarily home workers who did piecemeal weaving and other hand work which was in the process of being displaced by automation. In this case, it was a matter of large mills being set up with machinery that allowed much more work to be done with many fewer people. Their existing livelihoods were threatened and they rebelled -- destroying equipment and damaging factories until they were brought under control by continuously increasing violence on the part of the shop owners and government. This was the foundation of displacement by automation -- local retraining and dispersement of labor.

In the world today, a first phase of economic displacement has occurred because of unequal labor laws across the world. Some areas pay low wages in comparison to where the products are sold (sometimes the wages are not that bad in comparison to other wages local to the workers). Some bypass, or do not have, environmental laws or labor laws which prevent the more severe abuse of workers in the region and/or move the burden of cleaning the environment to the general society. In this first phase, the manufacturing base is still trying to do more with less; in this first phase, it is done by moving the work to places where they can make use of people who are valued less.

In the original Luddite period, automation came in and workers had to be retrained but the work and the labor remained local. Overall employment was (eventually) retained by retraining and by increasing the product market (increased consumerism and the rise of the "middle class"). In the first phase, production was moved to other locales to take advantage of the discrepancies of laws, working conditions, and attitudes toward general workers.

We are now entering the second phase. In this second phase, the cost of automation starts becoming less expensive, and more efficient, for the business owners than even that of outsourced labor and sub-component manufacture. In this second phase, it is still possible to have retraining but even with increased general education and increased consumerism it is not possible to fully compensate for the need of fewer workers per manufactured widget.

As the tip-over occurs, the global amount of unemployment will steadily rise. There are, of course, various options. Some philosophical/economic groups feel there is no obligation by society to care about the poor -- so those without employment are free to starve or scrape by in any way they can (except by attacking the privileged, of course). Other groups recognize this economic trend and have started to set up a stabilizing economic foundation so that, as automation continues to shift and replace the existing workforce, people have the ability to live at a sustenance (ability to house, feed, and educate themselves) level.

This concept of an economic foundation is called Universal Basic Income (UBI). It replaces various per-situation social supports (in the U.S., projects such as "welfare", "food stamps", "subsidized housing", and so forth) and allows for everyone to start off at a basic level no matter their circumstances. As they are able, and have opportunity, they add on top of this economic foundation to allow for "better" scales of living -- more space, fashionable clothes, less common foods, etc. The foundation exists for them to survive.

As is true for the various per-situation social supports, there are questions of need versus misuse. One of the concepts behind UBI is that, since **everyone** is entitled to the benefit, there is no actual potential of misuse. It does require a reorientation of attitudes of social responsibility. It has the very significant advantage of being a simple concept

-- although the actual implementation would require a huge shakeup of our economic structures.

It also requires answering basic questions that are difficult and controversial. Is the ability to survive a basic human right? There is no overall consensus on this question. Should the right to have an unlimited number of children be universal or is this an extra right that must be earned beyond the foundational level of sustenance? What are the minimal needs for survival? Food, shelter, heat (where needed), and minimal clothing can probably be agreed upon but what about items such as Internet access, cell phones? Is education a universal right? To what level might it be provided by the general society?

There are many other questions that must be resolved about UBI -- assuming that society decides that this is part of our overall social responsibility. There are social "experiments" in the process of being attempted right now and they should provide some ability to understand how various shifts change the economic interactions in society and the problems that surface.

95. Hunger and Appetite – The Lost Relationship – 7/22/2017

"Eat your vegetables." "Clean up your plate." "Don't you know that there are starving children in XXX and YYY who would love to have that food?" It may not be quite as true for children in the United States as it was 50 years ago, but I suspect many of the same statements are heard around the table. Of course, the last question of food distribution is more of a political issue than an economic or nutritional one.

On the positive side, I was not allowed to become a "picky" eater. If it was cold, I ate it. If it was spoiled, I ate it. If it was burned, I ate it. If I hated it, I ate it. (If I loved it, I ate it but might, or might not, have more.) I remember one time when I refused to eat something and I was required to stay at the dinner table until I cleaned my plate. As I recall, that was at about 4 in the morning. I possibly could have (and maybe not) "toughed it out" and "succeeded" in not eating it but I reached the point of realizing that I might win that battle but I would lose the ongoing struggle.

On the negative side, there was no control over portion size or what I ate. In fact, on those rare occasions when we ate out, I was not only expected to clean my plate but my mother seemed to feel that I was there as a garbage collector to clean her plate as well ("the growing boy needs his food and we are not going to waste any"). Any relationship between hunger and eating was burned out of me.

As a result, it is probably not a surprise that I had problems controlling my weight. My older brother was "blessed" with a more active metabolism and didn't gain weight until after his metabolism changed when he was about 30. He stayed skinny. I did not. It wasn't until I was away from home and had direct control over my eating that I succeeded (with the help of a swim class and a mile of swimming per day) in getting my weight under control.

That success did not carry over to the basics of appetite and portion control. I still had no correlation between how much my body needed and how much I ate. So, over the years, I regained the weight. During this same period (from the 1960s on), portion sizes at restaurants started on their continual gradient (see my "Supersizing" blog about the economics behind that trend) -- and "fast foods" became more of a daily reliance than an occasional treat.

The relationship between need and consumption is still not encouraged by the economic system of the U.S. Consumption is increased by various methods: commercials and general media as well as additives within the food. Addition of sugar adds to the calories, reduces the nutritional balance, and takes advantage of the "swallow reflex" to cause the body to want to eat beyond any (now ignored) response of fullness. Extra salt and/or MSG is added to enhance the other flavors to make it more appetizing without the need for careful preparation or more expensive ingredients.

It is possible to escape the cycle of eating without need (in the areas of the world where food is not in shortage) but it is "swimming up stream". First thing to do is to set portion control. I now have the mantra of "to waste or to waist" -- meaning that I may waste extra food but it is not going to add to the inches around my waist. It is still difficult for me to actually throw food away so leftovers come back to the house (adding to landfill contents and, often, still being thrown away when there is no opportunity to actually need it for a meal). And people in the U.S. (not such a problem elsewhere) expect larger portions for the prices that are paid.

After portion control (which can be done by mathematics rather than recovering one's ability to feel full) comes a deliberate reduction (or attempted elimination) of refined sugar and carbohydrates in the diet as well as the artificial sweeteners and substitutes which are, most likely, no better for your body (and possibly a lot worse). Not eating out gives much more control but processed foods are more accessible and more economic than home-prepared foods. I can buy grated cheese for less than I can buy block cheese and grate it myself (economics of scale, packaging, and distribution). In other words, a significant change in lifestyle as well as ignoring the economic, commercial, and media pressure to do otherwise is needed. Very difficult.

Another item which can help is to take time with meals -- both in preparation and in consumption. My wife talks about the 10 minutes that we get to spend with our sons at the dinner table before they head back off. In many countries, and societies, that is a half-hour or more. The average amount of work hours in the United States is so high that various other non-work activities are cut down. And one of the dominant areas of reduction is in meal time.

I am still fighting (at almost 60) to regain a relationship between need and consumption -- hunger and satiation. I make some progress but it is not easy -- society has evolved to make it undesirable for it to be easy. When you reach for the next bite of food, start asking yourself (this will slow down your eating also) whether you really want it. You may be surprised at the answer.

96. Equal Opportunity – That Fragile Illusion – 8/5/2017

Most people think that everyone should have equal opportunities -- opportunity for financial success, opportunity for professional success, opportunity to be and do what they want, opportunity for happiness (in whatever way it is defined). There are those few who feel that they can have what they want only by denying others -- but I believe that most feel that equal opportunity is a good goal.

There are also people who state that they believe that this is reality -- that there truly is equal opportunity for everyone. If someone is poor, it is because they just haven't worked hard enough. If someone doesn't have health care, it is because it wasn't really important to them to achieve. If someone has not achieved the professional and career level that they want, it is because they aren't good enough and/or haven't worked hard enough.

As is often true of perceptions, there is a bit of reality mixed in. All the above COULD be true -- maybe the person did not work hard enough, maybe they weren't good enough, maybe something wasn't important enough for them to choose -- they chose something else of equal cost that was of more immediate importance for them. But the idea of equal opportunity is that each person has EQUAL access to the opportunities that allows them to succeed in whatever definition that may be.

As I said earlier, most people think this should be and there are some who feel that that is the way that it is. It is a fragile illusion and is just not true for most people. Obstacles exist and not everyone has the same obstacles, or the same amount of obstacles, or the same ease of working through obstacles. While it is still possible for someone with many, many, obstacles to achieve "success" they may need to work many times harder AND encounter rare opportunities that many in their situation do not encounter (that is, they have to have "luck") as opposed to another person who has to work very little and have many opportunities presented.

At heart, there are a number of different areas which present these uneven obstacles. Although I will present them in categories, they might fit into multiple categories or be "properly" classified in a totally different category. I won't argue about such -- what I am presenting is the idea that, because of all of these obstacles which are not the same for all people, equal opportunity is an illusion and, in order for the chance of "success", sometimes outside help is needed. Some of these areas include:

- **Nutrition.** We presently have young children in Flint, Michigan growing up having had excess levels of lead in their drinking water. This will affect them all their lives. With continuously deteriorating infrastructure, including water lines and treatment plants, this is not unique to Flint even though it presently gets the headlines.

  Although few starve in the United States, malnutrition still exists (see previous blogs on nutrition and economics) because good nutrition requires good food and time to prepare and the poor often do not invest in this. With malnutrition, the full individual potential is difficult to achieve.

- **Income.** Income facilitates many things. It allows access to better schools and teachers. It allows more time to be spent on activities for learning and relaxation rather than for survival. It provides access to networks of people who also have easy access to facilities and opportunities which can make a huge difference.

  Even though general attributes which may lead to success exist independent of wealth, or pigmentation, or ethnic history, income does help, and always has helped, to eliminate obstacles and ease the road to opportunities.

- **Prejudices.** Opportunities can be explicitly denied to others because of the prejudices of those with the power. While prejudices can also exist on the part of the less powerful, those with power have the ability to "close the door" to others. Prejudices may be from most recent nationality ("Irish need not Apply", "No Chinese allowed", "Wetbacks not admitted", ...). They may be based on "race" (usually applied based on skin pigmentation, speech, or body patterns). They can be based on gender (male, female, trans-sexual, etc.). They can be based on religion, or ethnic background, or any other aspect that can be used as a separating label.

  Whatever prejudice exists, it can be used by those with power to limit opportunity and often has, both in the past and in the present.

- **Health.** "People have diabetes because they choose a bad lifestyle." Sorry, but I did not choose my mother or my grandfather. Perhaps I could have eaten, or lived, in a manner that would have lessened my chance to get diabetes -- but I have no control over my genetic history, the preservatives, hormones, pesticides, insecticides, food additives, or other aspects of our environment which are now leading to rising cases of diabetes.

  People do not choose to be born blind, or deaf, or without use of some of their limbs. While there are many strong, courageous people who have overcome such obstacles -- they are still obstacles that others do NOT have to combat and, thus, prevent equal opportunity.

  In addition to birth conditions and genetically-inclined diseases, people encounter other health issues through their lives. With proper health care, most can be worked with, but many people in the U.S. do NOT have full access to proper health care. Some health issues can arise out of personal choices (smoking, drinking, illegal and legal drugs, ...) and some feel that others should not help support them in the consequences from these deliberate choices. But there are many other choices that are NOT deliberate and even the deliberate bad choices are made from an environment of unequal possibilities.

- **Appearance.** OK. It shouldn't be this way but a taller person is more likely to be elected to office. A more fit person is more likely to be offered an opportunity that interacts with the general public. Societal views of "attractiveness" gives an advantage, or disadvantage, in the getting of jobs and of the likelihood of getting raises.

- **Access.** I talked about the aspect of income of giving access to opportunities. However, there are also physical access aspects. Currently, there is debate about getting rural access to broadband Internet -- it costs more to provide such but there is reluctance to charge rural people more (similar to the issues involving landline telephone access in the early and mid 20th century).

  If you live an hour from the closest library and cultural centers, it will make a difference. If your local town, or neighborhood, has limited possible choices and you cannot go elsewhere, it will make a difference.

- **Education.** Beginning with the time of Benjamin Franklin, and before, the idea of public education has been an important one to improve the chances of equal opportunities. Funding has always been a handicap to providing equal education but the struggle has continued with parents and teachers advocating for better, and more equal, public education.

  The "successful" -- even those who have been able to make use of private education -- have equally made use of public education to provide the workers needed for them to achieve their goals. Public education is needed to provide for the various types of workers needed within a society. Debate exists about how far along financial help should be provided for education -- but educated people are needed by society and, in my opinion, society should pay for what society needs. It is certainly an area to debate but it is true that the rich cannot exist without public education even if they do not directly make use of it.

- **Role Models**. Sometimes the use of language for various positions is made fun of as being "politically correct". However, if a postal carrier is always referred to as a mailman then people WILL think of a "man" in the role -- even though the gender has nothing to do with it. Similar to a congressperson. It would be possible to refer to people as congressmen and congresswomen but, in reality, gender has nothing to do with the role and it is both easier, and more accurate, to refer to them as congresspeople.

  On the visual side, if people see people who come from similar ethnic or religious heritage in a particular role, they will think that that is a role in which they might find themselves at some time. If a CEO that is pictured is a woman as often as a man, then the term will become non-genderized. People find it much easier to imagine themselves in a specific role if others who remind themselves of themselves can be seen in those roles.

- **Physical Environment.** A home is much more than a house, or apartment -- but having someplace stable to call home is important to people. It is important to their sense of stability and of being able to make plans for the future. In a similar fashion, it needs to be a safe place -- no rats coming out of the toilet or patches of plaster falling off the ceiling in the middle of the night. It needs to be sufficiently warm in winter and bearable in summer so that they can study, think, and relax.

- **Home Environment.** A final (for this blog) category of obstacle is that of the home environment. This is in addition to the physical environment. Are there people to talk to? People to encourage and help? People to comfort and aid when things don't go wrong? Is there an environment of angry survival or hopeful loving comfort and support? While there is not that much, in isolation, that society can do to help in this -- it CAN be helped with the removal, or minimizing, of the other obstacles that people encounter.

These are just a few types of obstacles that can stand in a person's way to the path of opportunities. What others come to mind for you? What are the best ways to make obstacles more even and allow something closer to "equal opportunity"?

97.   The Fragility of Life – 8/12/2017

Growing up, I was not in the very bottom of the U.S.'s income classes but, as discussed in my blog on Democratic class structures, we were in the lower income region. This basically meant that we survived but we never quite knew how we were going to survive. Each day was a recognition that my father might not have a job after the end of that day, my mother might be heading out again to find yet another service job, and that it might even mean moving once again. But, in comparison to many, I was still lucky and blissfully ignorant of how bad it might be -- I always had food, shelter, clothing and, although I was aware of those who did not have such, since I had always had them, I still mostly felt that I would continue to have them.

This stays with me, on a daily basis, as part of my individual psyche. Each day when I go to work, I recognize that I might not have a job when I come back. When I spend money I am concerned as to whether I am saving enough -- do I REALLY need that item -- or should it go into savings? But, if I put it into investments or savings, is it really safe? Should I, instead, live fully with my income as it is and just live with the reality that it might go away at any moment?

Life is insecure. Life is fragile.

My children haven't faced that environment and it is both good and bad. They don't have the fears and they do have a feeling of security. On the other hand, it is hard for them to really understand how others can be so concerned about their ability to live into the future. Empathy is also more difficult when you have always had store-made shoes in which to walk.

Life is fragile. It can be fragile because you are aware that a job is something that not everyone has and which you might not have after the end of the day. Some groups face something even more directly tangible -- you might be dead because someone feeling anger, fear, or hatred decides to kill you as you walk along the street.

Recognizing that life may end at a moment's notice affects the way people live. You live each day as best as you can -- or you give up (either option is possible with, in addition, complex mixing of both). You embrace family while, at the same time, you cherish them each day because either you, or they, may be lost before tomorrow. Or you reject family -- afraid of the pain that will occur if you lose them. You may live life loudly -- shouting instead of speaking -- out of joy of life. Or you may retreat into your own world drawing up the blanket to hope that the angels of death and misery will pass you by. Everything may be met with a laugh, perhaps your best shield against the pain and fear. Or there may be a blank face to the world trying to never give offense, never be noticed.

In regions of the world where war and poverty are more prevalent on a daily basis, life is seen as ghostly. You don't name your children until they are one or two years old because the chances are good they won't live that long. Material objects become meaningless (or everything) because a bomb may take them away as you are sleeping. Knowledge becomes more important because, most of the time, that cannot easily be taken. At the same time, the ability to survive may be paramount and leave no time or energy for seeking out knowledge that is not able to be immediately put to use.

Daily awareness, conscious or subconscious, of life's fragility molds the way many of us live our lives on a day to day basis. We may fight for greater opportunities and equality of treatment. Or we may take the other end of the teeter-totter and hold tight to what we have and denying others out of fear of loss of our own -- the fear of scarcity. We may look around and decide to blame others for our situation. Or we may look around and recognize what we do have and embrace that.

As difficult as being aware of life's fragility may make life -- lack of awareness, in my opinion, is much worse. No fear for the future because it exists as a tantamount right and anything done to stay on top of the mountain is as natural as water running downstream. Everyone else deserves to be in an inferior, lesser, position, because the divine right of ownership and leadership belongs to oneself. If you have the power to do what will benefit you then you use it for such. More is never enough because you are entitled to it all.

I don't have any decent arguments that indicate that reincarnation is a reality -- but it is something in which I WANT to believe. Each of us is placed in life with certain obstacles (even being rich can be an obstacle depending on the definition of the goal) and we struggle forward. Wouldn't it be nice to have further chances with newer, possibly less difficult, obstacles as we give it another try?

98.   Pay and the Perception of Fairness – 8/25/2017

Once upon a time, I worked for a high-tech division (Bell Laboratories) of a company. Within the company, the technical people knew what the pay scale was on a statistical basis. Each level of promotion was paid the square root of two, on average, times the average pay of the level below. Thus, my department head (the manager of my manager) would, on average, be paid twice the amount that someone at my level, on average, would get paid. This went up at least four levels. It did NOT go to the very top (more on that later) but it did go up a way.

There were sub-levels to which the square root of two did not apply. For example, the equivalent of a SW Engineer I versus a SW Engineer II did not connect to the ratio of the square root of two. But, in general, people knew what type of average deltas existed between levels (and even sub-levels, to a point). I keep saying "on average" because, as was (and still is) true for most companies, the monetary amount for each level was not publicly discussed -- neither the average nor the range.

This worked well. The precise amount did not matter too much. Maybe each level could have increased by the square root of three rather than two. It couldn't be a lot more than that, however, and still be considered "fair". It was known, and understood, that each level had more responsibility than the previous but -- beyond a certain ratio -- pay ratios could get out of whack (no longer at a reasonable point). Perception of fairness depended on two things -- visibility of ratios and a translation of that to a delta of responsibilities and duties.

How high up in the hierarchy did this pay ratio between levels apply? I don't know. As stated earlier, it made it four levels up. Beyond that, I don't know except that it did NOT apply to the highest executive levels. Thus, at some point, the ratios were discarded.

This discrepancy, or abandonment, of fairness at the higher layers was a topic of discussion. Most of the time it was just something in the background as we were doing OK within the company and system. One year, however, the CEO doubled their salary while doing a pay freeze for the rest of the employees. That did not go down well but it was known that most of the people were there because they could do the work they wanted to do -- not primarily for the money. But the perception of fairness never returned and morale never quite made it back to the same levels.

So, how should a CEO be paid? What is fair? If you used the formula I mentioned at the beginning here, then a CEO of a huge corporation should be paid about 16 times that of a regular worker. Is that close to the reality? Nope. Not even in Poland, where a CEO gets about 28 times that of a regular worker. If we used the (straining the "fairness" doctrine) square root of three (rather than two) ratio, it would mean that a CEO of a huge corporation would be paid 81 times that of a regular worker.

What is it like in the United States? About 354 times that of a regular worker -- it continues to go up and up. The highest in Europe is a ratio of about 148 to 1 in Switzerland. An article with more numbers is here.

So, once again, what is fair? More important, what meets the PERCEPTION of fair? Perception of fairness goes hand in hand with perceived mobility. If I can work hard and get that golden apple, that brass ring of the merry-go-round, that round-the-clock money mill then almost anything can be perceived as fair. Dreams of winning the lottery just don't allow reality and statistics to interfere.

Note that there are two categories of CEOs -- those who are heads of private corporations and those who are heads of public corporations. The CEOs of private corporations often have relatively low salaries because they have large amounts of equity (ownership) and recognize that putting money back into the company (if successful) will give them a higher eventual rate of return. Taking out a lower salary allows the money to be used to actively grow the company. When I was "Vice President of Engineering" of a company that I co-founded, my salary was only a little (not even the square root of two) higher than the rest of the employees.

The CEOs of public corporations have salaries, perks, "golden parachutes", and so forth determined by the board of directors. They will have equity as part of their compensation but their salaries have no direct interaction with the equity. In other words, a lower salary does not translate into higher potential equity and a higher salary does not mean less equity.

So, how is the compensation package of the CEO of a for-profit (non-profit is similar but generally has a ceiling beyond which people won't tolerate) public corporation determined? First, please note that public corporations (those which have had a successful Initial Public Offering (IPO)) are going to be beyond the small scale of most private corporations. Saying, as an example, that an average employee's compensation package (salary, taxes, benefits, physical overhead, etc.) is $50,000/year this would indicate an average U.S. CEO salary of $17.7 Million/year. Obviously, a company grossing $15 Million/year cannot pay this. Probably a company would have to start reaching a $500 Million/year gross income before their board of directors would even consider a $17.7 Million/year compensation package for the CEO.

The above referenced study actually indicates the average CEO salary in the U.S. is approximately $12.3 Million/year (less than that $17.7 Million/year of the example calculation). I have not seen a distribution graph of CEO salaries. However, the highest paid CEO in the U.S. for 2017 is $236 Million/year. My suspicion (not verified by any hard data) is that the distribution does not follow a standard bell curve. I think that there are a lot of lower salaries from smaller companies that bring down the average from the higher paid ones. But, since the average still exists at 354 times average worker, there also has to be a lot of very highly paid ones.

These highly paid CEOs are not really in the salaried employee category. They are Rock Stars, Professional Sports

First Draft Picks, Celebrity Panelists, and so forth. Their compensation packages are based on competition between boards of directors which keep inflating compensating packages in order to "win" -- show that they have a better CEO because they are paid more. There is no tie between company performance and package and most have such extensive separation packages that someone can be hired as a CEO, do a horrible job, and leave with enough money to support 1,000 (or more) families for a year.

Boards of directors are theoretically determined by the stockholders -- but stockholders care primarily about quarterly results and, with the mega corporations, even effective losses of hundreds of millions of dollars, don't affect the bottom line much. If that is so, does it really matter what the CEOs are paid? Some say no, it doesn't. Personally, I think (like the situation from my earlier career days) it poisons morale. And, whether it affects the stockholder's value or not, that $236 Million/year is paid for by at least $1000/employee in the one company. In other companies, it may affect the "average" employee even more. But it is still all part of the package of valuation of employees at companies -- not just the CEO's compensation package.

99.   Economic Interconnectivity and Big Data – 9/23/2017

The world economy is a huge set of interconnections. One type of job depends on other types of jobs; if a job type disappears it is likely to affect many other job positions. Scarcity of resources of one type can affect the prices of many cascading products. If the world does shift from fossil fuels to renewable energy sources, new jobs will appear and old ones will change or go away.

The interconnectivity also causes great fragility as the world gets larger and there are more dependencies. Imagine, if you can, people waking up tomorrow and deciding that the Internet is no longer of interest (I can easily remember when it didn't exist) -- how many products would no longer have a market, how many people would no longer have a job, how would it affect others (advertising, for example -- and printed newspapers might surge back into dominance)?

Once upon a time, I was interviewing with Google and, as part of the telephone interview, we discussed potential projects and interests. I put forth the idea that, since Google was well designed to integrate knowledge and had such massive data storage and access, they would be well able to create an economic model of interconnected occupations and salaries. At this point in time, I would like to also add in products and localized market prices.

Why bother with any type of tool? Why not just make the change and see what happens? The main advantage of such a tool is to have a better ability to forecast the effects of policy changes. What really happens if minimum wage is increased to a living wage? What happens if the illegal immigrants who are largely responsible for hand harvesting of our fruits and vegetables are kept away -- what will be the effects on produce prices, truckers, grocery stores, and so forth? What jobs are affected if private transportation is minimized and public transportation maximized?

Such a project would be impossible if every individual, unique job, discrete part, and location had to be tracked. Luckily, items can be aggregated -- 500 Blue F-150 trucks should only have a quantity value change over 1 Blue F-150 truck (but, at the same time, there needs to be a way of describing Red F-150 trucks without having a fully different item). There is a lot of work to be done and it would still be a difficult project but certainly within the capabilities of many of the larger data handling companies -- Google, Facebook, Amazon, IBM, Microsoft, ... What would be the Return On Investment (ROI) for such a project? It's really difficult to know but it would be a valuable service/project that should be of use to governments and businesses around the world.

I would suggest architecting such a project as an iterative accretion of data. Start with something relatively small -- a loaf of bread. The loaf of bread has a set of occupations associated with it -- baker, packers, delivery people, stockers, advertising, payroll, Human Resources, etc. It also has a set of ingredients -- flour, yeast, filtered water, possibly milk, salt, and so forth. Each ingredient has an amount which acts as a ratio of strength in the links to the bread. Each ingredient has its own delivery and production chain which each have associated costs and value. It would be considerable in itself but the greatest value would be the fact that it is still small enough to be thrown away. New links and new data structure values will be discovered to be needed as the database develops. Now do it over (iterate) with those better values and links. Do it again if needed. Now add butter to the bread and continue on.

There are also usability concerns. The bread company may start off selling only white bread and then add rye bread -- each with their own percentage of sales. How does one substitute recipe ingredients? How do you change the dependencies and the ingredient ratios? What happens if a problem ruins the rye crop for the year? If modelling an auto, how easy is it to change the model from gasoline to electric? Not only is there a substitution of an "ingredient" but the interconnections to suppliers, dealers, raw materials (batteries, possibly lithium) change. The model must be able to be changed easily because modelling the existing situation may be interesting but comparisons are what gives the most value.

How would you address such a problem? What do you see as specific practical benefits from such an economic model? Is there some subset of such a model already in existence that could be used as the core of expansion? How are unpaid people incorporated into the model, recognizing that the system falls apart without them -- even if they are not considered to be part of the Gross Domestic Product (GDP) or a paid occupation?

While I find the project fascinating just from a theoretical basis, I keep finding more and more potential uses as I consider the matter.

100.   Hocus Pocus: The Distance Between Magic to Science – 10/21/2017

Arthur C. Clarke, famous science fiction and science fact writer, once said that *"Any sufficiently advanced technology is indistinguishable from magic".*

Although stories told about witches primarily exist from what is, in the West, often called "the Dark Ages", the underlying problem continues to exist. Someone -- especially a social loner who is doing something not considered "normal" -- has knowledge and experience beyond that of others. If it is useful, they may be allowed to exist to be used by others. That is, they allow it to be used until something bad happens and they use the target of their fear and ignorance as a scapegoat and attack them.

Hanging on my waist, I have a device that has computational power much greater that that of a 1951/52 UNIVAC that occupied an entire room 65 years ago. If you took that same smartphone and presented it in the town of Salem, Massachusetts 325 years ago, you could look forward to being on the non-preferred side of the Witch Trials. However, if you took the smartphone back to 1952, most of the technology would be totally unexplainable. Depending on your audience, they would declare it to be an amazing fake, a stage device, or -- yes, magic. Very few would believe that it was real but you most likely would not be subject to being burned at a stake.

Science consists of building upon previously discovered knowledge. Any jump in knowledge is typically met with suspicion. Einstein's theory of general relativity was a jump in understanding. It took years to be accepted and the "in-between" steps are still being proven even unto this day (2017 Nobel Prize being given for detection of "gravitational waves"). Leonardo Da Vinci was sufficiently wise to keep most of his ideas and discoveries isolated within his journals. His public face was largely concerned with his works of art for the Church and the rich. Since it didn't happen, we cannot know for certain, but I suspect that if he had succeeded in building, and demonstrating, a functional flying machine it would have had, at best, very mixed reactions from the Church and public.

The split in perception between science and magic works both directions. Something that would be considered "commonplace" within current society (even if really understood by only a small subsection of the people) would be considered "magic" in the past. When people envision things in the future, it is often classified into "science fiction" UNLESS it is some ability or behavior that does not have an obvious basis in current science. Flying cars are science fiction. Functional "spells" are magic.

Current, scientifically acceptable, spells are called algorithms. They piece together various simple instructions into logical frameworks and decision networks and come out with a "game App" or a "streaming video App" or a "communal workspace App". All of these would have been considered magic in the past. I will make the guess that there are a lot of things, about which we speculate as magic, that will also become commonplace in the future. In order for applications (Apps) to work, however, the convenient microcomputer/smartphone must also be present. Will it be true that, in order for Harry Potter's spells to become valid that some other foundation device must be created?

What ideas of the future would you classify as magic? Do you see scientific paths to have them realized? What ideas would be the most inexplicable if sent to the past?

www.ingramcontent.com/pod-product-compliance
Lightning Source LLC
Chambersburg PA
CBHW032148080426
42735CB00008B/627